THE SCIENCE OF MARVEL

From Infinity Stones to Iron Man's Armor,

the Real Science Behind the MCU Revealed!

Sebastian Alvarado, PhD

ADAMS MEDIA
NEW YORK LONDON TORONTO SYDNEY NEW DELHI

A **adams**media

Adams Media
An Imprint of Simon & Schuster, Inc.
57 Littlefield Street
Avon, Massachusetts 02322

First Adams Media trade paperback edition April 2019

ADAMS MEDIA and colophon are trademarks of Simon & Schuster.

For information about special discounts for bulk purchases, please contact Simon & Schuster Special Sales at 1-866-506-1949 or business@simonandschuster.com.

The Simon & Schuster Speakers Bureau can bring authors to your live event. For more information or to book an event contact the Simon & Schuster Speakers Bureau at 1-866-248-3049 or visit our website at www.simonspeakers.com.

Interior design by Michelle Kelly
Interior images © 123RF/ahasoft2000; Getty Images/gn8, bubaone

Manufactured in the United States of America

10 9 8 7 6 5 4 3 2 1

Library of Congress Cataloging-in-Publication Data
Names: Alvarado, Sebastian, author.
Title: The science of Marvel / Sebastian Alvarado, PhD.
Description: Avon, Massachusetts: Adams Media, 2019.
Includes index.
Identifiers: LCCN 2018055601 | ISBN 9781507209981 (pb) | ISBN 9781507209998 (ebook)
Subjects: LCSH: Comic strip characters--Miscellanea. | Marvel Comics Group--Miscellanea. | Science in popular culture. | Technology in popular culture. | BISAC: SCIENCE / Astronomy. | SCIENCE / Physics. | PERFORMING ARTS / Film & Video / General.
Classification: LCC PN6714 .A42 2019 | DDC 741.5/9--dc23
LC record available at https://lccn.loc.gov/2018055601

ISBN 978-1-5072-0998-1
ISBN 978-1-5072-0999-8 (ebook)

CONTENTS

CHAPTER 4: FANTASTIC PHYSIOLOGY   67

CHAPTER 5: GLORIOUS GADGETS   93

CHAPTER 6: AGGRESSIVE ASSAULTS   117

CHAPTER 7: MECHANICAL MARVELS   137

CHAPTER 8: STUPENDOUS SCENARIOS   155

CHAPTER 9: FANTASTIC PHYSICS   173

CHAPTER 10: MAGNIFICENT MARVELS   197

DEDICATION

To the Marvelous Maral Tajerian and our three little experiments, Minas, Flo, and Maxine.

INTRODUCTION

How does a blind Daredevil perceive the world around him? How does Spider-Man use protein purification to make spider silks for his web-shooters? What kinds of physical forces are bouncing around Thor's head when he gets struck by lightning while wielding Mjolnir? If you've wondered about these and other questions from the Marvel Universe, you're not alone.

What you may not have realized is that these sorts of questions are a gateway to different scientific fields of study.

From this book you'll learn that science can sometimes be as fantastic as fiction. Each of the forty-three entries will describe scenes from the Marvel Cinematic Universe. You'll also find speculations for what could be happening in the scene and real-world technology that corresponds to Marvel's imagined science.

And you'll see that a lot of that science isn't imagined—it's *real*! The writers and artists who created Marvel's lore always drew inspirations from the world around them. In the 1950s and 1960s, for example, people were terrified of the atomic bomb. That hysteria, boosted by the Cold War, fueled stories about radioactive spider bites or blasts of gamma rays. The results? Spider-Man and the Hulk, among others. As people learned more about genetics, the X-Men showed that mutant genes formed the basis of incredible powers. In Marvel's world, physics, genetics, and chemistry provided a framework in which science could be used to enhance humans, not threaten them.

So, true believers, join me as we explore and explain the uncanny, the incredible, and the amazing world of Marvel.

Chapter 1

Bustling Brains

HAWKEYE'S AIM

WHEN: *Thor, The Avengers, Avengers: Age of Ultron, Captain America: Civil War*

WHO: Hawkeye

SCIENCE CONCEPTS: Psychology, visual attention, physics of archery

INTRODUCTION

Expert marksmanship and most athletic feats require rigorous training over many years. In archery, this involves developing the strength to pull a bowstring in order to deliver the power necessary for an arrow to cover large distances and penetrate a given target. This strength must be shaped throughout the body and its posture so it can work like a well-oiled machine that shoots an arrow with accuracy and precision. Archers, like any expert athletes, must also train their minds to coordinate their gaze on stationary or moving targets. How does a normal human such as Marvel's Clint Barton achieve such skill to become the Avenger known as Hawkeye?

BACKSTORY

Throughout multiple appearances in the MCU, Hawkeye is a S.H.I.E.L.D. operative with the uncanny ability to never miss his target. While his apparent skill is seen with all sorts of firearms, his use of bow and arrow affords him the stealth he needs to carry out covert missions as well as the utility to reuse ammo. With his bow he can execute a series of standard shots by shifting his balance and stance and several trick shots behind his back at long range. He is able to fall off a building, shoot a grappling arrow straight into concrete, and hit many targets he isn't even looking at. Needless to say it's pretty impressive for someone who has no superpowers.

THE SCIENCE OF MARVEL

To understand the talents of Clint Barton it's important to understand how humans learn a skill that requires mastery of hand-eye coordination as well as being able to shoot a firearm or arrow when needed. To achieve this proficiency as a S.H.I.E.L.D. agent, Barton must have spent several years doing drills and exercises that involve taking down enemies in different scenarios. These training regimens provided him with enhanced perception in circumstances that require combat; they primed him to make decisions very quickly and anticipate the actions of his adversaries. While his feats may look easy on the big screen, it's likely he has been training his mind and body for his entire life.

Throughout his many appearances, Hawkeye uses both compound and recurve bows to take out his targets. The main difference between these two types of bows lies in the amount of strength required to pull back on the bowstring while preparing to shoot an arrow. As you draw the string on a recurve bow, it becomes harder and harder to pull. In contrast, drawing the string on a compound bow starts off hard but "lets off" once it's fully drawn. This is due to the pulleys at the ends of the bow that help relieve the stress on the bowstring. This let-off allows Hawkeye to take more time to make his shot without succumbing to muscle fatigue and would be most useful for stealth attacks where his target is unaware of his presence (for example, when he is asked to take out Thor in *Thor*). Regardless of the type of bow he's using, though, he requires a reasonable amount of upper-body strength: after all, his arrows are capable of penetrating robots and concrete (probably requiring more power as the distance between him and his target increases!).

Once the bowstring is pulled, what ensures Barton makes his shot? To consider this we have to follow the target as he perceives it. Consider the typical pathway of light through the eye before it becomes coded into the neural firing processed by the brain (see The Human Lie Detector). As light depicting a given scene enters Barton's eye, it passes through a tear film into the curved cornea, which focuses light past the iris (the colored part of your eye) into the pupil. The light passes through the lens where it is further focused onto the inner layer of the eye, the retina.

As Barton finds his target, his eyes converge, his lenses focus, and his pupils constrict. If his target is far away, the ciliary muscles, which hold the lens in his eye, need to relax, causing the lens to flatten. In contrast, closer targets require the ciliary muscles to contract, making the lens rounder and allowing him to focus on images closer to him. Since Barton needs to make adjustments frequently and very quickly in battle, these ciliary muscles are under constant strain, and the lens constantly changes shape. Hawkeye is likely past his forties, suggesting that the lenses in his eyes are not nearly as flexible and able to focus on targets as they were when he was younger, regardless of how hard the ciliary muscles are working. Earlier in life, he probably received intraocular lens implants made of silicon, foreseeing (pun intended) that his eyes and his aim are his most valuable assets as a S.H.I.E.L.D. operative.

THE SCIENCE OF REAL LIFE

It wouldn't be fair to compare Hawkeye to an Olympic archer since these archers spend their training hitting a stationary target. Hawkeye is hardly afforded that luxury. Regardless, it should be noted that Hawkeye has inconsistent form when anchoring his draw hand onto his chin. For this, his bowstring needs to be aligned close to his line of sight in order for his brain to process accurate aiming. Any variance in this process forces his brain to guess more often than employ any skill in hitting a target. These poses may have also been decisions by the film's director to allow audiences to see the actor's intense face as he is in the heat of battle (I see you, Jeremy Renner). Additionally, Hawkeye makes several trick shots in which he hits a target he is not even looking at. While these feats could be the result of incredible luck, some incredibly difficult shots can be learned. Consider the Parthian shot, a tactic employed by nomadic mounted archers in ancient Eurasia. As enemies pursued a small number of retreating cavalry, they discovered that these mounted soldiers were trained to shoot pursuing targets. They turned backward on galloping horses, while steering with only the pressure of their legs (no stirrups!).

Outside of technical form, there is a certain depth to how visual attention works when we compare skilled athletes to an average person. Dr. Joan Vickers, a kinesiologist at the University of Calgary in Canada, was the first scientist to describe the phenomenon of "steady" or "quiet" eye in golfers. She found that better players maintained a longer and steadier gaze on the ball before, during, and after striking it. These findings were replicated in a variety of other sports such as football, basketball, and (of course) archery. In this 500–3,000 millisecond pause, experts take 62 percent longer than untrained individuals to slow their thinking, decelerate their heartbeat, and decrease their physiological arousal to achieve a flow state. While skilled athletes naturally pick up this gaze behavior, trainers have also been able to use quiet eye training to make better athletes. In a 2001 study, Vickers recruited three university basketball teams over two seasons. One team received quiet eye training, including gaze fixation on the hoop and backboard during drills. Over two seasons, the quiet eye–trained group significantly improved its shooting accuracy over the two control teams.

While quiet eye training can affect the outcomes of a given motor skill, it has been suggested that it also can contribute to an athlete's ability to perform under pressure. In work led by Dr. Samuel Vine at the University of Exeter in England, participants were asked to do 520 basketball free throws over eight days. Participants were given quiet eye training or technical training and then subjected to anxiety-inducing conditions. Like stern high school gym coaches, the researchers told participants that they had performed so poorly in the exercise that they landed in the bottom third of their cohort, possibly disqualifying them from the study. Stressed-out after receiving this information, the control group lost accuracy; however, the quiet eye trained participants did not. While the exact neural mechanisms underlying quiet eye duration still require some study, this behavior may offer insight into how an average human such as Clint Barton never misses a target.

MANTIS'S EMPATHY

WHEN: *Guardians of the Galaxy Vol. 2, Avengers: Infinity War*
WHO: Mantis
SCIENCE CONCEPTS: Empathy, cognition, neuroscience

INTRODUCTION
When you relate to another person you use various social cues to approach the situation. If an individual is laughing, perhaps you'll enter the conversation smiling. If that same person is crying, you may give them privacy. Most humans follow some predictable social cues in these types of situations. These responses may also vary depending on your own emotional state, feelings, and personal background. For example, if you've had a terrible day, someone else's laughter in the room may not prompt you to enter the conversation with a smile. Our empathy in these social situations allows us to understand the thoughts and emotions of the people around us.

BACKSTORY
In *Guardians of the Galaxy Vol. 2*, we are introduced to the insectoid alien Mantis. Mantis, having been adopted by Ego, a Celestial, knows very little about social interactions with other life-forms. For example, she often misses the meanings of jokes, figures of speech, and how to avoid making others feel awkward. Interestingly, what she lacks in social aptitude, she makes up for with her abilities to empathize with any living organisms through touch and her glowing antennae. This ability offers Mantis a view into the emotional state

of others; thus, she is able to alter their emotions. She has used these powers to bring out Peter Quill's love for Gamora as well as being able to subdue powerful beings such as Ego or Thanos.

THE SCIENCE OF MARVEL

Mantis appears a bit insect-like with a pair of antennae that act like a sensory organ, analogous to those of Earth's insects. Her powers stem from the unique way in which her antennae glow when she makes contact with someone she is trying to empathize with. They may be detecting chemosensory information in the environment that helps her better understand an emotional scene. They are not directly needed for her abilities, but they supplement her empathic abilities through an added sense of smell, wiring through sensory neurons into the olfactory bulb of her brain. The glow of her antennae may even be similar to the biochemistry of bioluminescent reactions seen in Earth's insects, which are often used for social communication.

Given the requirement for touch, it's possible that Mantis is detecting emotional states through electric skin conductance. In humans, electric skin conductance changes as we go through various states of arousal (broadly defined as emotional states). This response can be measured in the palms of your hands and soles of your feet where you have a high density of eccrine sweat glands. These glands secrete water to the surface of your skin, causing it to cool, and are directly wired to your sympathetic nervous system, which is ruled by the limbic parts of your brain. In other words, you have little control over this part of your physiology, and it can telegraph emotional states in very subtle ways. If Mantis can detect these small electric currents and smell trace pheromones with her antennae, perhaps she is capable of creating a composite image of a person's emotional state.

For example, in one scene of *Guardians of the Galaxy Vol. 2* Drax shares his sorrow with Mantis regarding his wife and daughter, murdered by order of Thanos. In that moment he has visited a very traumatic memory, signaling to his limbic system to alter the secretion of sweat from eccrine glands in his skin. This immediately results in a drop in skin conductance followed by evaporation of a unique chemical pattern of pheromones from sweat glands all over his body. Mantis's antennae detect a layer of his pain when trace amounts of these pheromones bind to receptors in her antennae. Her empathy is further refined by feeling with her touch the drop in skin conductance. This information converges on the supramarginal gyrus of the cerebral cortex, sitting at the junction between the parietal, frontal, and temporal lobes of her brain. This area of the brain has been described as responsible for distinguishing our emotions from those of other people. While unable to extract details of Drax's memories, she experiences his pain in a profound way. Despite being unable to understand social cues given by language or gestures, she understands Drax in a way he's never been understood before.

THE SCIENCE OF REAL LIFE

In real life, we are particularly efficient at understanding various social signals and empathizing with our peers. We have evolved elaborate parental care strategies that allow us to empathize with several social cues our child may present. These can range from the sound of a cry to the furrow of a brow. All of these characteristics register with a part of our brain and help us understand the needs of our baby, allowing us to be better parents who can tend to the child's needs. Humans have leveraged our ability to empathize to improve a group's fitness and cooperate with one another. In other words, we indirectly benefit ourselves by caring for our peers through empathic processes. This behavior is not exclusive to humans. In a study led by Dr. Jules Masserman at Northwestern University, a rhesus macaque monkey was willing to starve itself for twelve days rather than press a lever that would shock a peer.

Outside of this visual and auditory information, we even employ our sense of smell to understand certain behavioral cues. Consider a study led by Dr. Shani Gelstein at the Weizmann Institute of

Science, examining behavioral responses to human tears. She collected negative-mood tears from women who watched a sad movie, *The Boxer*, in a secluded room. While male participants in the study were unable to differentiate by smell between the real tears and a saline solution, they reported lowered sexual arousal after smelling the tears, which was also accompanied with lowered salivary testosterone. In a separate study, the same research group found differences in brain activity between the men who sniffed negative-mood tears and those who did not. Researchers traced the differences to areas of the brain commonly associated with sexual arousal such as the thalamus. This sense of smell isn't only evident with tears; it also takes advantage of the many smells that are volatized (that is, they come from our skin as vapors) while we sweat in various emotional states. In a study out of Rice University led by Dr. Denise Chen, smelling the sweat of men and women who were subjected to an anxiety-causing stimulus (watching scary movies such as *The Shining*) can improve cognitive performance.

Our ability to empathize is processed in two ways, cognitively and emotionally. This cognitive component requires a certain level of detachment to consider the emotions of another person, whereas the emotional component requires a reaction to the emotional state observed. Currently our best understanding of empathy has relied heavily on brain imaging studies that have linked different parts of the brain to empathy in different scenarios. Research led by Dr. Tania Singer at the Max Planck Institute in Germany investigated the emotional aspect of empathy and how our own emotional state can bias our empathy. Using pleasant and unpleasant visual-tactile information, pairs of participants were asked to evaluate the emotions of their partners. During these trials, activity within the right supramarginal gyrus (a distinct fold on the cerebral cortex) was required to autocorrect our own egocentricity. That is, our innate reaction is to think our own emotions come first. Using transcranial magnetic stimulation this group was able to disrupt the activity of this part of the brain, causing participants to be more likely to project their own emotions onto their partner.

MEMORY ERASURE

WHEN: *Captain America: The Winter Soldier, Captain America: Civil War*

WHO: Winter Soldier

SCIENCE CONCEPTS: Learning, memory, neuroscience

INTRODUCTION

Over the course of your life, you will take part in a multitude of experiences. You may form influential social networks, learn about the natural world around you, and contribute to this world through your own legacy. These life experiences and how they influence your actions will define your identity in a variety of ways. If you were to trace the abstract idea of your identity to a physical state, it would probably take the form of behavioral patterns and memories stored in the form of connections between neurons in your brain. Over the last several decades, we have been able to reveal a great deal about how these connections allow us to store and retrieve information and even how to ablate this process. In this regard, can we reveal how to erase memory or even implant new memories as seen in Bucky Barnes's transformation into the Winter Soldier?

BACKSTORY

In *Captain America: The Winter Soldier* and *Captain America: Civil War*, Bucky Barnes is subjected to brainwashing methods by HYDRA scientist Arnim Zola. During these procedures, Zola is able to erase Barnes's memories and train him to follow HYDRA's orders. Over time Barnes loses his identity and any memory of his past self or beliefs, making him the Winter Soldier. The effects of this procedure are so profound that, under orders from HYDRA agents, Barnes is willing to knife, strangle, and rocket his childhood BFF, Steve Rogers. Barnes's brainwashing eventually wears off during the events of *Captain America: Civil War* and during his rehabilitation in the Kingdom of Wakanda.

THE SCIENCE OF MARVEL

In the MCU, Bucky Barnes's transformation into the Winter Soldier requires a complete erasure of his identity in order to rebuild him as a weapon. As depicted in *Captain America: The Winter Soldier*, Bucky Barnes is subjected to repeated sessions of electroconvulsive therapy to erase his memories. In these sessions electrodes are placed onto his temples, and electrical pulses are run through his skull, initiating a storm of neuronal firing throughout his brain that causes seizures. Following electroconvulsive therapy, the frontal and temporal lobes undergo the most prominent changes in activity. These areas of the brain control a wide range of behaviors such as motor function, problem-solving, memory, language, judgment, impulse control, and social and sexual behavior. Over time, these parts of the brain decrease their neural activity and cerebral blood flow and create a brain-wide trauma that changes Barnes's stress axis and dopamine system. In addition to the disrupted electrical firing of neurons in the brain, a secondary cascade of stress hormones and dopamine likely flooded his synapses, causing long-term rewiring of his brain and cognitive processes.

To complement electroconvulsive therapy, Barnes may also have been subjected to drug-induced amnesia. Drugs such as alcohol and benzodiazepines, when taken in combination and in high doses, could prevent the formation of long-term memories. This approach would have been most effective in erasing the memory of a recent mission as opposed to eliminating a biographical recall of Barnes's previous life. Drug-induced amnesia may also double as a means to induce a coma, which would prepare the Winter Soldier for cryostasis. While most humans are likely to overdose with these treatments, the Super Soldier Serum that Barnes received could have possibly provided him with the cerebral fortitude to withstand extreme doses. Arnim Zola probably saw this resilience as an opportunity to push the limits of these pharmacological avenues that would potentially kill an average human. After losing his memories and identity, Barnes was probably susceptible to thought reform while held in captivity.

While it may be hard to believe Barnes could become a ruthless killer, we should note that his training as an American infantry soldier is only moderately different than his role as the Winter Soldier. Before becoming the Winter Soldier, Barnes was already trained to obey a chain of command, use various firearms, and fight in close quarters combat. HYDRA just realigned his belief system to change whom he received his orders from. This process of thought reform started with isolating Barnes from the life erased from his memories. Any reminder of his past could consolidate memories that defined his previous identity, interfering with HYDRA's reprogramming. Isolation further provided control over all of Barnes's behavioral patterns, causing dependency on his captors. During this process, Zola may have also seeded doubts about Barnes's loyalty to his friends, family, and country—and theirs to him—while reinforcing repetitive patterns of behaviors aligned with HYDRA beliefs. This sense of doubt would force him to question his own role in a larger narrative shaped by Arnim Zola, even allowing the HYDRA agent to implant false memories through social conditioning.

THE SCIENCE OF REAL LIFE

To understand how Bucky Barnes was able to lose his memories and identity, it will help to first unpack how a memory would typically have formed in Bucky's brain before his transformation. Consider a salient memory shared by Steve Rogers in *Captain America: Civil War*. Rogers reminisces about a time during their childhood where they had to ride in the back of a freezer truck because both had wasted their money on hot dogs and impressing girls. Each time either Steve or Bucky recall this memory, the memory itself changes. At a cellular level, neuronal networks that fire recalling ideas like "freezer truck," "hot dog," or "money" begin to integrate, strengthening his memory. Neurons within these networks strengthen their firing by changing the number of receptors that interface chemical communication between neurons. Soon enough this memory becomes a formative one that reinforces Barnes's social connection with a friend like Steve Rogers. If you were to erase this memory, you would have to reverse this process at the molecular and cellular level.

While not entirely backed by peer-reviewed publications (possibly due to ethical concerns related to conducting relevant experiments), behavioral control through "thought reform" can be used to influence patterns of behavior. Behavior can be forced through compliance (regardless of how it aligns with core personal beliefs) or persuasion (which involves changing core belief systems). Across history, we have drawn on similarities in how nations and cults have shaped human belief systems. These processes often employ an agent who is given total control over a subject and his or her behavioral patterns (when to sleep, eat, go to the bathroom, etc.). The purpose of these exercises is to break down the subject's identity to the point that he or she can be introduced to a new one that aligns with the tenets of a new belief system. This process is often accentuated by social influence within a community and peer pressure and the threat of social exclusion or possibly even danger. For example, many POWs in the Korean War were considered brainwashed by enemy forces, but in reality these POWs were isolated, in fear of their captors, and subject to torture (compliance). Of the more than seven thousand US prisoners captured during the Korean War, only twenty-one actually refused to return to the US, citing a true change in their belief system (persuasion). As expected, rescued POWs were quick to reassume their identity and place in their society once they returned.

There are other subtle ways in which memories can be modified or even implanted due to the way in which the recall of one memory can change the memory or even tie it to a different memory. In a seminal study carried out by Dr. Elizabeth Loftus at the University of Washington, she succeeded in implanting novel memories that had actually never happened. In this study, sometimes referred to as the lost-in-the-mall experiment, Loftus recruited pairs of participants. One member was designated the subject, and the other was an older sibling or parent. Subjects were mailed a booklet with four stories about them that occurred between the ages of four and six, with one of these stories being a complete fabrication. The false story would "recount" an event in which the subject, at around the age of five, was lost in a large shopping mall for an extended time and eventually helped by an elderly woman and reunited with her or his family. After weeks of reading and revising this booklet, 37 percent of participants incorporated a false memory into their own history, one that never happened.

MIND CONTROL

WHEN: *Avengers: Age of Ultron, Captain America: Civil War, Avengers: Infinity War*

WHO: Erik Selvig, Hawkeye, Scarlet Witch, Loki (in possession of the Mind Stone)

SCIENCE CONCEPTS: Transcranial magnetic stimulation, optogenetics

INTRODUCTION

Every time you have a thought, there is always some physical, chemical, and physiological underlying process. Whether we have a full understanding of that process is still not entirely clear. We know that your brain has cells that communicate with one another using a combination of chemical and electrical signals. If we had complete control over the firing of a single neuron, its circuit, or a pattern of circuits, we could possibly regulate behavior with great precision. So given the technology and the knowledge we have now, how easy is it to manipulate behavior? Can we control minds the way that Scarlet Witch and the Mind Stone do in the MCU?

BACKSTORY

In the MCU the ability to control minds has been used by and on various Avengers. The earliest example of this ability was Loki, who was in possession of the Mind Stone inside his scepter; he used this to control Hawkeye and Erik Selvig in *The Avengers*. Similarly, Wolfgang von Strucker used the Mind Stone to imbue Wanda Maximoff (Scarlet Witch) with the power to telekinetically move objects and control minds. Using these powers she single-handedly turned the Hulk against the Avengers and made Black Widow, Captain America, Iron Man, and Thor face their worst nightmares as hallucinations in *Avengers: Age of Ultron*. What physical forces would allow the control of someone's thoughts and hallucinations?

THE SCIENCE OF MARVEL

Since Maximoff's powers were derived from the Mind Stone, we can assume her abilities are a diluted form of what the Mind Stone is capable of. This makes sense when you consider that the Mind Stone is capable of controlling a subject for days, whereas the effects of Maximoff's powers are short-lived and more mood-altering. Given Scarlet Witch's comic book parentage and the nature of many of the Infinity Stones, I'd guess her abilities involve generating powerful and localized magnetic fields through the movement of her hands. This is how the noninvasive procedure of transcranial magnetic stimulation (TMS) can be used to stimulate or block activity in different parts of the brain. When we consider the functions of electricity and magnetism we know we are actually discussing two different forms of the same phenomenon. An electric current run through a coil will create magnetic fields that can define the polarity of a magnet. Reciprocally, a strong magnetic field can also modulate an electric current. During TMS, a strong electromagnetic force is targeted toward the surface of the brain to depolarize neurons in that area, leading to increased activity.

Imagine, as Scarlet Witch signals an okay sign with her thumb and index finger she creates a "coil" that conducts enough current to create a localized magnetic field completely under her control. This coil can be used to electromagnetically create a current through the scalp and skull of her target, inducing an oppositely directed current in neural populations within the brain. Maximoff likely has a PhD-level understanding of neurology, since her intuitive abilities make clear she has an advanced working knowledge of the human brain, its neuroanatomy, and its function. This know-how enables her to differentiate her control over inhibitory and excitatory neurons in different parts of the brain, allowing her to elicit a range of behaviors. This power could permit added control if she could shrink the area of effect of such a magnetic field to a single neuron!

If her ability to generate superconductive coils of electromagnetic fields acts like conventional TMS rigs, she may have access to cortical layers on the surface of the brain at depths of 3–6 millimeters. However, a stronger magnetic field would allow for deeper stimulation. During Tony Stark's visual hallucination, she may have sent a wave into the subcortical layer and into his limbic system, causing the release of neurotransmitters and triggering his PTSD. This hallucination may have also sent several other rippling effects to other parts of his brain, forming connections between auditory, visual, and somatosensory cortices and consolidating the perception of the hallucination being real even though it is not. Given Maximoff's stealth approach to her victims, she may even be using her powers to implant visions in the occipital lobe where vision is processed; this part of the brain is located in the back of the head (see The Human Lie Detector). Thankfully, Maximoff takes a turn toward good, since the ability to control minds paired with telekinesis would truly make her the most powerful Avenger on the team.

THE SCIENCE OF REAL LIFE

At a noninvasive level, we are able to manipulate several kinds of behavior, using different technologies. The speculative use of transcranial magnetic stimulation can in fact regulate the neural firing on the cortical layers of the human brain. Dr. Alexander Kendl and Dr. Joseph Peer of the University of Innsbruck in Austria have even postulated the idea that lightning discharges may generate magnetic fields similar to those used in TMS. In their calculations they hypothesized that individuals within 20–200 meters of a lightning strike may be subject to visual hallucinations that look like glowing spheres of light or the rare phenomenon of "ball lightning." At a clinical level, Dr. Ian Cook at the Semel Institute for Neuroscience and Human Behavior at UCLA uses TMS to treat depression by eliciting a rewiring of neurons. TMS is clinically approved only for treating depression but it also has been shown to reduce chronic pain, migraines, and anxiety in research studies. While the technology is capable of modulating emotional states and moods, it lacks the refined and immediate control over executive function seen in the MCU.

Outside of its clinical uses, TMS is a particularly powerful tool to help understand how the brain functions in a living human. In a study led by Dr. Rebecca Saxe at MIT, researchers were able to use TMS to modify human moral judgment. In Saxe's earlier work, she identified the temporoparietal junction (TPJ) as the area of the brain that would activate when participants were asked to evaluate someone else's intentions. To validate this finding, she designed a study where individuals were read stories that involved one character doing something objectively wrong to the other. After listening to the story, study participants were asked to rate the behavior of the villain of the story on a scale of 1 to 7 (1 = absolutely forbidden and 7 = absolutely permissible). Study participants robustly identified the malicious culprit with the highest score except when TMS was used to manipulate their TPJ. This type of study would have been impossible to carry out without tools that can manipulate function in the brain.

At the most invasive level, we can even use genetic modification to control behavior in the brain, using light-sensitive channels that can trigger neuronal firing. This approach, called optogenetics, was pioneered by Dr. Karl Deisseroth of Stanford University. Using this technique, Deisseroth was able to repurpose light-sensitive channel proteins from algae and transgenically introduce them into neural populations of a mouse's brain. When activated by light, these channel proteins generate a change in the electric potential of a cell's membrane, causing a neuron to fire. Using transgenic mice or viruses we can introduce these algal proteins into specific parts of the brain and remotely control neural circuits using a fiber optic cable implanted into the brain. This tool has been extensively used in the field of neuroscience to understand which neural circuits underlie behaviors related to hunger, sexual arousal, thirst, memory, and more. Furthermore, a variety of these channel proteins can be tuned to different wavelengths of light, which makes it possible for researchers to induce a repertoire of firing states in neurons. While optogenetics is most often used for basic research questions, it has shown promise in potential treatments for restoring sight in the blind.

Chapter 2

Curious Critters

CELESTIAL ASSIMILATION

WHEN: *Guardians of the Galaxy Vol. 2*

WHO: Ego, Star-Lord

SCIENCE CONCEPTS: Genetic organization, hybridization, parthenogenesis

INTRODUCTION

In the very first instant you existed as a single-cell embryo, you had the mixed genetic information of your mother and father. A few seconds before that moment, genetic material from sperm and ovum performed an elaborate dance to reorganize the living blueprint that makes you. In order to do so, these cells obey a series of rules that guarantee a healthy start to life. Sometimes, however, those rules can be bent. For example, what happens when these genetic blueprints don't come from the same species? In what other ways can new genes be transferred between genomes? Using Ego's biology as a guide, we will try to figure out how Ego was able to hybridize with so many different species.

BACKSTORY

In the events of *Guardians of the Galaxy Vol. 2*, the Guardians find themselves confronted with a Celestial, Ego. These Celestials are among the oldest and most powerful humanoid beings in the universe with unique abilities. Unlike any other organism in the universe, Ego has the ability to manipulate the matter around him to create anything, from a fancy fountain to an entire planetoid chassis for his brain. Specifically, he is capable of creating proxy bodies of himself that look like different alien races but carry the hereditary information of a Celestial. He uses these proxies to reproduce with life-forms from other planets in order to assimilate the universe. However, he is unable to successfully pass on his Celestial powers to the next generation until he meets

Peter Quill. How exactly can Ego hybridize his DNA with so many different species?

THE SCIENCE OF MARVEL

Following the events from *Guardians of the Galaxy Vol. 2*, it seems particularly important for Ego to propagate his Celestial powers through sexual reproduction. He sets up a universe-wide genetic cross using his own genome in order to individually screen that population for another Celestial. Whenever he didn't find a Celestial hybrid, he sacrificed that child and repeated the experiment. Assuming that the function of genetics isn't only earthbound but literally universal, Ego would need to integrate his genes into his offspring the good old-fashioned way. However, he has several hurdles in his way to hybridizing with different species. If he can't obey the same molecular rules that endemic species follow, how did Peter Quill become the only successfully Celestial result from the giant genetics experiment Ego carried out?

Humans (and most plants and animals that sexually reproduce), rely on male/female sexes and their respective vectors for genetic material (sperm/egg) to reproduce. The evolution of sex is in part driven by the advantage of diversifying genetics within a population and a larger palette of biological traits that can be selected for. However, for this to work properly, a critical process has to work smoothly during that moment fertilization happens between Ego and Meredith Quill. The instant sperm and egg fuse, the ova's twenty-three chromosomes are partly into metaphase and about to align along the center of the cell. The sperm loses its tail and coat and begins to decondense its set of twenty-three chromosomes. Both sets of chromosomes replicate and prepare to be paired up and given to daughter cells. This all raises an important question: how would this work in other species with forty, thirty-two, or 200 chromosomes? Consider: some species of roundworms can have two chromosomes, while species of hermit crabs can have up to 254 chromosomes. In such a case, Ego would have to reshape his genome and its architecture across the protein scaffolds that build chromosomes each time he finds a mate. This complexity would impede his plans and partly explain the difficulty of siring an individual who has the right Celestial genes to "connect with the light." Also, if he knows exactly which genes are needed to make a Celestial he would have gotten it right the first time, right?

So perhaps Ego is experimenting with the genetic material he is contributing to his various kin. If this is the case, it would be far too taxing to line up synthetic chromosomes of exact size and structure with those of a human (or anything else). In such a predicament, he may be giving his genetic material in small bits and pieces through lateral gene transfer. In this case, instead of sperm he could be a pseudo-virus, removing parts of his genome and infecting the ova of Meredith Quill. Eventually, he hopes, the right parts of his genome will sustain a connection to Celestial power in the right child. However, this plan still wouldn't fertilize the egg (it would have only half the number of chromosomes it needs for cell replication). The only way an unfertilized egg (infected by Ego's genes) could form an embryo would be through the process of parthenogenesis. This relies on using mitosis (cell division) to create sex cells with a full complement of chromosomes and can be induced in mammals with electrical and chemical stimuli. This process in the MCU may explain why Peter Quill is more human than Celestial in appearance ("Congratulations Ms. Quill, it's a floating brain!"). This is a stretch, but then again we *are* talking about mating a metaphysical brain-planet with a human female.

THE SCIENCE OF REAL LIFE

In real life, we do get forms of hybridization that can occur between two closely related species (usually within the same genus). For example, a mule can be bred from a donkey (sixty-two chromosomes) and a horse (sixty-four chromosomes). In this case, there is enough genetic similarity across chromosomes for some pairing and successive cell divisions to occur. However, the cells of this resulting mule will not have the genetic constitution (with only sixty-three chromosomes) to make its own sex cells; that means mules can't breed. These sex cells require an even number of chromosomes to pair up properly with a mate and lead to fertilization. Furthermore, the differences between parental animals will be large enough to impede the generation of viable sperm or ova in their offspring.

During the process of hybridization there is also the important contribution of genomic imprinting during fertilization. Genomic imprinting relies on the dosing of paternal or maternal genes in the resulting offspring. For example, in the chromosomes you carry from your mother and father, some eighty different genes are imprinted with instructions to activate only the mother's copy of a gene while silencing the father's copy of the same gene (and vice versa). Many of these imprinted genes are critical for pre- and post-natal development. A good way to illustrate this is to compare the hybrid of a father lion and mother tiger (liger) and the hybrid of a mother lion and father tiger (tigon). While each crossbreed has different paternal/maternal species, the ways in which parental genes become imprinted are different. Smaller tigons look more like their tiger fathers and larger ligers look more like their lion mothers. In most of these cases, the hybrid becomes sterile.

RARE OFFSPRING

In rare circumstances, a hybrid such as a mule can give birth to a foal. In a case from Collbran, Colorado, genetic testing of the mule-delivered foal revealed the genetic link between mother and daughter. However, there was no genetic information from the mule's paternal donkey in the foal, suggesting that the mule's reproductive biology excluded the paternal genome before undergoing a form of parthenogenesis (that is, the development of an egg without fertilization).

Parthenogenesis has evolved in multiple species across the animal kingdom such as insects, reptiles, and sharks. While parthenogenesis has never been observed to occur naturally in mammals, it can be induced in ova through calcium immersion/injection and electric pulses. This induction has been carried out in various lab animals including mice, rabbits, and even monkeys. As of 2007, we also have been able to induce parthenogenesis in human ova. In work led by Dr. Elena Revazova of the International Stem Cell Corporation, the first parthenogenetic ova were generated in hopes of creating a novel source of donor-matched embryonic stem cells. The discredited scientist Hwang Woo-suk had in fact made a parthenogenetic embryo in the experiments in which he claimed to have collected stem cells from a cloned human embryo. In reality he had not cloned an embryo, but he had induced parthenogenesis among sex cells. In mammals, parthenogenesis can be carried out only in females since the size of egg cells makes them large enough to carry enough nutrients to sustain subsequent cell divisions; sperm cells are too small. This would suggest that instead of a Peter Quill as Star-Lord, a Petra Quill as Star-Lady might be more realistic.

GIANT ANTS

WHEN: Ant-Man, Ant-Man and the Wasp
WHO: Ant-Man, Yellowjacket
SCIENCE CONCEPTS: Developmental growth, DNA methylation, giant arthropods

INTRODUCTION

Insects have varied dramatically in their size throughout their natural history. Although it might elicit shudders in the squeamish, the fossil record shows that 315 million years ago ancestors of today's dragonflies were about the size of a pigeon and millipedes were 2 meters long and half a meter wide. What drives these dramatic changes in insect size over evolutionary time? Here we will discuss what allowed some insects to grow and some to shrink.

BACKSTORY

Throughout *Ant-Man*, different species of ants are enlarged to assail anyone who stands in the way of Henry Pym, Hope van Dyne, and Scott Lang. In the first *Ant-Man* movie, one bullet ant is accidentally enlarged in a faceoff between Yellowjacket and Ant-Man and then kept as a house pet by Cassie Lang. This same ant later serves as a life-sized decoy for Scott Lang during his house arrest in *Ant-Man and the Wasp*. Additionally, in *Ant-Man and the Wasp*, several enlarged ants are used to guard Henry Pym when he's tied up by Elihas Starr and Ghost. These giant ants are helpful allies in a fight and also double as manual labor for moving around heavy machinery for Henry Pym's projects.

THE SCIENCE OF MARVEL

Ants in the Marvel Cinematic Universe are useful in their miniature size, either serving as stealth agents to block security cameras, hide secret plans, or transport the miniaturized Ant-Man suit behind bars. However, some ants are handpicked from Henry Pym's myrmecological infantry for deployment as enlarged versions. Throughout *Ant-Man* and its sequel *Ant-Man and the Wasp*, these ants are enlarged using blue Pym Particles. Upon enlargement, these ants increase to the size of a large dog, with the strength to lift more than a thousand times their weight.

Considering the physiology of most insects, such an increase in size would put a large demand on their metabolism to respire enough oxygen in order to produce the energy needed to move. A bigger body requires more energy, more respiration, and more oxygen. This is an easier feat for a human since vertebrate mammals have efficient ways to intake oxygen to meet our energy needs. For example, we can increase our breathing rate and our lungs have a large internal surface area that actively traffics oxygen through a closed circulatory system. Despite this, Scott Lang could only maintain his Giant-Man size for a few minutes and still had to sleep for three days to recoup his energy. Ants, unfortunately, are not as efficient when receiving oxygen and rely on the passive diffusion of ambient oxygen into small openings on their abdomens called spiracles. Air flows into these spiracles and diffuses across their cell membranes into their hemolymph (bug blood), providing fuel to the muscles and allowing the insect to locomote. Although these ants are scaled directly from their small sizes, the respiratory benefits they enjoy as tiny insects would not transfer over to the larger versions. This means that these ants would have to modify their behaviors to increase oxygen circulation into their spiracles through flexing their abdomen to "pant." Perhaps Henry Pym has these ants moving in well-ventilated areas in his lab, which would *slightly* offset the respiratory cost for oxygen. Otherwise, these ants would have to carry out anaerobic respiration that would shorten their life living large. Considering available colony numbers this could be plausibly (and wastefully) sustained provided there are enough Pym Particles to enlarge more ants when the already enlarged ants died of suffocation.

Assuming the respiratory demands of these ants are met, some types would definitely be go-to species for enlargement because of their predatory prowess. In *Ant-Man and the Wasp*, the species kept in Henry Pym's Altoids container appear to be trap-jaw ants. These ants have mandibles that function like a bear trap and use them to capture and kill prey. These jaws can close at speeds of 35–64 meters a second, making the trap-jaw ant's attack one of the fastest self-powered strikes in nature. In their MCU enlarged state, these jaws would be about twice the length of a rolling pin and exert enough force to chomp off a limb. When not used for a predatory strike, these mandibles also exert recoil force that can propel the ant's body away from danger. In the wild, the tiny version of these ants use the recoil of their bite to escape from antlions.

THE SCIENCE OF REAL LIFE

Did ants ever get as big as they did in *Ant-Man*? Not quite, but our fossil record reveals that fifty million years ago some ants were as large as a hummingbird. Before ants evolved, we had even larger insects during the late Carboniferous and early Permian periods (about three hundred million years ago). During this time, atmospheric oxygen was at an all-time high (35 percent compared to today's 20 percent). Vast swamp forests were photosynthesizing more gaseous carbon dioxide and releasing more oxygen into the atmosphere and bacteria had not yet evolved to decompose plant detritus that would release carbon dioxide back into the atmosphere. These two factors pushed atmospheric oxygen to high enough levels to influence the ecology of insects, allowing them to meet their respiratory needs for oxygen at a larger size.

Today, evolution has curbed these sizable ambitions, causing a selective pressure for smaller insects. Entomologists propose that this could have been the result of changes to atmospheric oxygen and the fact that insects during their developmental stages were easier prey at larger sizes. Insects that metamorphose like ants develop as fatty larval instars that pupate in a cocoon, emerging as fully grown adults. Giant larvae and pupae would serve as a very nutritious and easy-to-catch prey, a vulnerability that posed too much of a threat to the survival of the species. Over time natural selection curated the ants that became small enough to maintain a colony and also small enough that larger predators would find them to be not worth the trouble. Currently, one of the largest species we know of, the bullet ant (native to Central and South America), grows to about 4 cm.

Is it also possible to enlarge an ant in real life? From personal experience, I would say yes. During my graduate training at McGill University in Canada, I was interested in the underlying molecular processes that determine size variation in animals. I was lucky enough to work with myrmecologists (ant people) who maintained colonies of carpenter ants. While it was easy to identify the largest and smallest individuals in a colony, we were also interested in the graded variation of their size. In my research, I measured a chemical modification, called methylation, that gets painted onto DNA during the ants' development. This molecular mark acts like an off switch for gene function. We found that it was generally higher in larvae that developed into workers versus those that developed into larger soldiers. We then applied drugs that could decrease and increase the amounts of DNA methylation and were able to nearly double the ants' size and even generate individuals smaller than those that exist in nature (who doesn't need smaller ants?). We eventually mapped some of this change in methylation to a key regulator of growth, and we found that a graded amount of methylation on this gene was actually able to drive its expression along a continuum. That is, a 10 percent increase in DNA methylation at a single site of this growth regulator could affect a 10 percent change in size. So between two discrete sizes in the colony, we were able to paint many shades of gray that corresponded to a continuum of variation. This was a pretty novel finding since it provides new understanding about how a chemical off switch can generate natural variation in a trait such as size.

I AM GROOT

WHEN: Guardians of the Galaxy, Guardians of the Galaxy Vol. 2, Avengers: Infinity War

WHO: Groot

SCIENCE CONCEPTS: Plant physiology, plant form and function

INTRODUCTION

Plant life has evolved some of the most incredible adaptations over the course of Earth's history, making plants among the most successful organisms in the world. They are one of the only organisms that can harness the sun's energy to sustain not only their own growth and development but that of all animal life.

BACKSTORY

One of the most lovable and relatable characters in the Marvel Cinematic Universe is everyone's favorite botanical buddy, Groot. This sentient humanoid walking tree of the species *Flora colossus* is capable of using various aspects of plant growth to his advantage for both defense and offense. He can become impervious to gunfire by growing a thick husk of bramble or project long vines from his arms to attack enemies. When not directly involved in combat, he can change the shape and length of his body at will and regenerate lost parts of his body. Lastly, due to a stiff larynx made out of wood, everything he apparently says sounds like "I am Groot." However, those who know him well detect the nuances of his speech and understand him much better than less familiar acquaintances.

THE SCIENCE OF MARVEL

If we are to assume that Groot is an analogous form to plant-life from Earth, we can deduce several interesting facts about his unique powers. His most amazing abilities stem from (pun intended) his incredibly fast cell growth and division. This ability allows him to pick and choose the tissues in his body to grow and specialize in a variety of ways. These tissues are probably derived from an analog of cell pools in various parts of his body capable of rapid division (meristem). These meristematic cells remain undifferentiated and tightly packed next to each other with small chloroplasts and thin cell walls maintained in a balance between cell renewal and organ initiation. A cross-section of Groot's arm would show a ring of this meristematic tissue as the layer beneath the bark that generates the vasculature needed to transport water (xylem) or sugars (phloem). Additional outer layers of this tissue would also form the cork cambium that feeds dead tissues to the outer layer of Groot's bark-like skin where it becomes hard and protective. The ability of these cells to replenish lost tissue explains how Groot can grow a new arm when Gamora cuts it off in their first encounter in *Guardians of the Galaxy* or develop into baby Groot from a small branch. In this specific case, Groot allows himself to reproduce asexually through a pool of meristem cells in a leftover branch under the careful protection and care of Rocket.

How does Groot possibly sustain the necessary energy to carry out all the feats he does through rapid cell division and growth? Humans need to constantly respire, taking in oxygen to metabolize sugars and fats in our bodies and producing the energy needed for all of our movements. Plants, on the other hand, use their leaves to photosynthesize CO_2 into sugars to meet low-energy demands. Groot only has a few leaves on his crown and budding off as twigs on his extremities and wouldn't sustain the tissues he grows. Our best understanding of his metabolism has to consider what happens to trees on Earth that lose their leaves in the winter. This seasonal occurrence relies on a tree's ability to photosynthesize during the spring and summer months and store that energy in the bonds of sugars. While the specifics of Groot's metabolism are not overtly stated in the MCU, an end-credits scene in *Guardians of the Galaxy Vol. 2* may offer a hint; we see a young teenage Groot in a messy room and an entire floor filled with piles of dead leaves and Star-Lord berating him for his lack of cleanliness. This suggests that Groot may produce more leaves than we typically see when he is in motion or interacting with his teammates. Once he has metabolized enough sugars from ambient CO_2 in his ship and solar radiation he may even store these carbohydrates as resins or saps in his torso (making him even sweeter than he lets on). In essence, he could have reversed the typical metabolic need to photosynthesize for an Earth tree's version of dormancy in order to adopt a mobile life cycle.

THE SCIENCE OF REAL LIFE

Our hope to see trees walk is not just limited to the fictional Groot. *Socratea exorrhiza*, or the "walking palm," has been speculated since the 1980s to be capable of lifting itself up using its stilt-like roots and scuttling its trunk through the forests of Central and South America. Some locals have even claimed that the tree can move up to 20 meters in a given year. However, a 2005 report from the Center for Sustainable Development Studies in Costa Rica suggested that this is mostly a myth spread by local tour guides. Rather than making it possible for these palms to "walk," the stilt roots seem to provide additional support to the trunk of the tree, which allows the tree to grow taller without investing energy into increasing its girth. Otherwise, the ability to exhibit reactive movement appears to be restricted to plants such as *Mimosa pudica* (called "sensitive plant" because it folds up its leaves when touched) or *Dionaea muscipula* (Venus flytrap). In *Mimosa* plants, it is hypothesized that the folding response evolved as an herbivore deterrent by reducing the surface area available to eat and presenting a drooping and wilted appearance. In Venus flytraps, the plant's movements are tied to a means of collecting nutrients from insects in an environment where the soil is heavily depleted of nitrogenous resources that allow a plant to grow.

WORLD-RECORD GROWTH MOVEMENT

If we consider how growth can affect so-called movement, we can consider the fastest growing plant, which holds a Guinness world record for growing 91 cm a day: a bamboo. This rate would clock its growth-based movement at 0.00003 km/hour. A screen of gene expression led by Dr. Jian Gao of the International Center for Bamboo and Rattan in China has revealed an expected role for several plant hormones. Changes in gene expression provide a hint as to which synthetic pathways are more active in two types of closely related bamboo species that can grow fast and slow. Specifically, increased regulation of auxins and decreased regulation of abscisic acid suggest their involvement in the faster-growing shoots.

The means by which Groot conserves his energy don't stray too far away from how deciduous trees survive harsh seasons. This trait evolved along polar clines and allowed trees to survive harsh winters (similar to how many animals have evolved to hibernate). This change is initiated by changes in day length and temperature. Throughout spring and summer, the tree is actively growing, photosynthesizing and storing sugars in the phloem in the vascular bundles of its trunk. In the fall, green chlorophylls that facilitate photosynthesis begin to get broken down to be recycled in the coming spring. This leaves other pigments that are normally unseen, such as yellow and orange carotenoids, to become more evident. Depending on the cold temperatures that remain above freezing, additional anthocyanins that lend purple and reds to the foliage of the tree begin to be expressed. The synthesis of abscisic acid in the leaf stems causes leaves to fall and the tree enters dormancy with a significant drop in temperature. This is about the time that we collect maple syrup in maple trees!

ROCKET RACCOON

WHEN: Guardians of the Galaxy, Guardians of the Galaxy Vol. 2, Avengers: Infinity War	
WHO: Rocket	
SCIENCE CONCEPTS: Animal behavior, brain evolution, intelligence	

INTRODUCTION
Humans often like to think of ourselves as the cleverest animals on Earth. We've been incredibly successful because we can shape our environment around our needs, often by developing new technologies. Our ancestors' ability to craft tools, develop language, and pass information down through culture are all considered behavioral milestones that have allowed us to become the "intelligent"

animals we consider ourselves to be today. Underlying our human ingenuity, however, are anatomical changes that have enabled our intelligence, creativity, and resourcefulness to evolve. What selective pressures encourage the evolution of the brain and how it solves problems?

BACKSTORY

Despite looking exactly like a raccoon, 89P13 (better known as Rocket) is a particularly intelligent member of the Guardians of the Galaxy. In addition to being able to build or repair pretty much anything, he can pilot spaceships, fight expertly hand-to-claw, and has escaped twenty-three high-security intergalactic prisons. Central to all of these talents is his unrivaled resourcefulness. While in imprisoned in the Kyln, Rocket was able to improvise an escape for his team in less than five minutes by turning off the prison's artificial gravity and hijacking nearby security drones to propel a guard's watchtower to the docking bay. Similarly, when given more than five minutes, Rocket is capable of building some of the most powerful weapons and bombs the galaxy has ever seen. So what exactly happened on Halfworld that made Rocket so clever?

THE SCIENCE OF MARVEL

Rocket is "created" on the mysterious Halfworld where he is taken apart and reassembled through a series of cruel cybernetic and genetic experiments. The result of these torturous procedures uplifts Rocket to a bipedal mammal with the ability to speak and use advanced technology. In order to break down what makes Rocket so special, we will start by examining the neuroanatomical changes that may have coincided with his transition from rodent to cyborg and their underlying genetics. Additionally, we will discuss the nature of neural implants that Rocket may have had surgically implanted into his brain to improve the various cognitive processes he gained on Halfworld.

If we were to look deep within Rocket's brain, it would show increased volume in the frontal, parietal, and temporal lobes, as these areas are often correlated with intelligence. These parts make up part of our cerebral cortex, which is the outer layer of the brain made up primarily of gray matter. This gray matter is made up of neuronal cell bodies, their dendrites (the branching of neurons), and the connections between neurons (known as synapses). In contrast, the inner layers of the cerebral cortex are mostly made of white matter, or simply the axons connecting parts of the brain that function as a conducting track between different parts of the brain. The cortical thickness within these lobes is involved in several cognitive processes related to perception, language, memory, and consciousness, and it is likely to become thicker as Rocket's intellect grows.

In addition to cortical thickness, the presence of cortical folding is a hallmark of sentience and intelligence in higher vertebrates. These folds that we see on the surface of human brains increase the surface area of the brain and the working space for neurons to make new connections in gray matter. During embryological development, cell division and interactions within the physical limitations of the skull contribute to the formation of these folds. As the brain gets bigger, newly migrating neurons from the hypothalamic tracts create an outward pressure against the skull. Migrating neurons are capable of neatly organizing themselves on the surfaces of the cortex (as they do in the smooth brains of a rat or mouse) or they can stick to each other, causing clumps that can result in folds (as seen in humans). At a genetic level, this could be modified by turning off genes that regulate the stickiness between cells (FLRT1/3) or by turning on the genes that amplify the number of migrating neural progenitor cells along hypothalamic tracts (ARHGAP11B). Control over stickiness between cells and the sheer volume increases through cell division would make it possible to create more folds, more surface area, and more connections to support Rocket's intellectual growth.

At the cybernetic level, a neural implant could have facilitated Rocket's early development when he was a raccoon kit if reward circuits in his brain were stimulated during a training regimen that reinforced his advanced behaviors. These neural implants take advantage of a basic cellular property of neurons: their ability to conduct and produce an electrical current. Implants would be able to sense and administer an electrical current to initiate the neuromodulation needed to elicit certain sensations or emotions. We can pick from a number of reward and pain centers in the brain that would generate the needed response. If Rocket successfully solved a complex puzzle, for example, he could receive a pulse to the ventral tegmental area of the brain, an important reward center. If he were to make a mistake, stimulation of the dorsal posterior insula would result in an intense sensation of pain. Remote control over these implants during training would accelerate the rate at which Rocket could learn skills and develop more advanced cognitive capacities.

THE SCIENCE OF REAL LIFE

How smart is an actual raccoon compared to Rocket? It will probably be some time before a raccoon can fly a spaceship but they are nonetheless intelligent and resourceful creatures. Also, the processes of modifying genes and using brain implants are already widely practiced in neuroscience; however, the results are not a wisecracking mouse, rat, *or* raccoon.

What talents does an earthbound raccoon already possess? For starters, looking at the sheer number of neurons, a raccoon has about 450 million neurons packed in a cat-sized brain, making its neural density similar to that of a primate. When it comes to senses, a raccoon has one of the most developed senses of touch ever studied. This is in part due to a particularly large somatosensory cortex in its brain and a large number of sensory neurons for touch within its hands. Raccoons' hands possess sensitive "whiskers" called vibrissae, which allow them to sometimes identify an object before they touch it. Lastly, their feet can turn 180 degrees, providing them flexibility while climbing down a tree head first. Just considering the dexterity in their feet and their acute sense of touch, we can conclude that Rocket would make a better pilot than Star-Lord.

PROBLEM SOLVERS

Raccoons have a track record of solving problems and adapting to new environments. In studies carried out in the early 1900s, raccoons proved to be talented escape artists with an excellent memory. In these experiments, raccoons were confined in a box with a combination of different latches and buttons that needed to be pressed to be unlocked. The animals solved most of these puzzle boxes in fewer than ten tries and, after a whole year, still remembered their solutions.

Regarding the use of genetic modification, certain genes have been added to mouse models that change their ability to learn. For example, the FOXP2 gene has a role in how speech is formed in humans. When a human version of FOXP2 was introduced into a mouse line, it increased the frequency and complexity of their vocalizations and improved a mouse's ability to solve a maze. Similarly, neural implants can help alleviate the effects of brain damage to help an animal complete a task. In research carried out on rhesus macaque monkeys, individuals were trained for two years on a delayed match-to-sample task. This involved seeing a photo, experiencing a delay, and then touching a screen with seven pictures (one of which was a match to the first photo). Monkeys were given a neural implant that would detect and record the neural firing that would happen when they did the task correctly, allowing researchers to record a "shadow" of the brain's activity when it carried out the task correctly. Subjects were then administered cocaine to dull their decision-making abilities and performed worse on the task. However, when the neural implant was activated to reproduce the correct shadow of brain activity, the monkeys were able to again perform the task correctly.

CHAPTER 2: CURIOUS CRITTERS

Chapter 3

Nervous Neuroscience

THE HUMAN LIE DETECTOR

WHEN: *Daredevil, The Defenders*

WHO: Daredevil, Stick

SCIENCE CONCEPTS: Sensory neuroscience, psychology, brain processing

INTRODUCTION

As you turn the pages of this book your fingers have a tactile feel on the edges and corners of a page. Your ears can hear paper folding, sliding, and bending. Your eyes scan page numbers, titles, and paragraphs. Perhaps you can even smell the pages of the print stock that was used for this book. (Please don't *taste* the book.) You have a variety of senses that perceive your surroundings, get processed in your brain, and provide you the necessary information to choose your behaviors. Given these senses can you detect the intentions of another individual? What if we were to remove one of these senses, such as your ability to see?

BACKSTORY

In *Daredevil*, Matthew Murdock is blinded at a young age in a chemical spill while trying to save an elderly man. Murdock develops superhuman sensory perception in his other senses, with an increased ability to hear the world around him. This becomes a huge advantage as Murdock, studying to become a lawyer, can detect the intentions of his clients, prosecutors, and witnesses. As the Devil of Hell's Kitchen, he uses his new abilities to surveil the city by tuning out the noise of sirens and ambient noise to hone in on conversations spoken between gangsters. In a fight, Daredevil can battle in the dark and can't be surprised by his assailants.

THE SCIENCE OF MARVEL

Before his accident, Murdock saw the world in color and in motion. Photosensitive rods and cones at the back of his eye transmitted information presented as light and encoded it into electrical impulses that fired within axons bundled in his optic nerve. This information traveled through the center of his brain via the optic chiasm and lateral geniculate nucleus in order to be processed on different sides of his brain. Information received on his left visual field was transmitted to the right side of his brain toward the cortical layers of the occipital lobe at the back of his brain (and vice versa for information received in his right visual field). Once the electrical impulses reached the layers of his occipital lobe, this information was further processed through connections that run dorsally and ventrally over the cortex of the brain. However, the moment Matthew Murdock was sprayed with toxic chemicals, this entire process was destroyed. Over time the visual cortices in his occipital lobe began to atrophy, but they still retained some connections to other parts of his brain. In contrast, other somatosensory, language, and auditory cortices grew in thickness and formed new connections to the occipital lobe, connections that a sighted human does not have.

Soon enough, Murdock is capable of hearing a pin drop in a noisy room and locating it perfectly. Eventually his abilities reach the point where he can situate sounds through concrete and flesh. In order to pinpoint the sources of sounds, Murdock uses the same sensory processes we all use; his skills, though, have been improved significantly through repeated training and auditory exercises. When you hear a sound, it travels as a wave that compresses air toward the outer ear and into the ear canal, where it hits the eardrum. From the eardrum, information gets transmitted through three bones (ossicles) that connect to the spiral-shaped cochlea. Inside the cochlea, hair-like cells receive vibrations and encode them into the auditory nerve as electrical impulses, which signal this information to the temporal lobe of your brain. In a given auditory field, Daredevil uses the distance between his ears to situate the source of a sound using a sound wave's intensity, timing, and changes in its frequency.

> **TIMING, INTENSITY, AND FREQUENCY**
>
> Sound waves arrive at each ear at a different time, giving the brain a time lapse to process an auditory field in the horizontal plane. In a similar fashion, the intensity (or amplitude) of sound waves is perceived differently by each ear (a sound closer to one ear has a higher intensity). If a sound is coming from above, the outer ear changes the sound wave's frequency, causing your auditory cortex to process it within the vertical plane.

When he was a child, Daredevil's enhanced hearing ability initially manifested itself as an amplification of every noise in the room. While the frequency of noise is detected by the ear, the ability to eliminate ambient noises in order to hone in on the ones that matter is a cognitive process carried out in the brain. Through his training with Stick, Murdock tuned these sensory circuits in his auditory cortex by strengthening the connections between neurons that process auditory information. At the molecular level, the neurons that form these circuits increase the density of receptors that receive relevant neurotransmitters, thus increasing their sensitivity and poising them to process even the faintest of electrical impulses.

THE SCIENCE OF REAL LIFE
In real life, a blind person can't hear the heartbeat of another person across the room, but the brain does process auditory and tactile information differently in a blind person than in a sighted person. The earliest record of such increased cognitive function was recorded by the eighteenth-century French philosopher Denis Diderot. In his *Letter on the Blind for the Use of Those Who Can See*

he described how two blind men, one a Cambridge professor and another a winemaker, were capable of using their remaining senses to perceive the world in great depth. At the time, the blind were stigmatized and considered a burden to families, the church, and society, making Diderot's letter an "eye-opening" account of the cognitive abilities of the blind. Interestingly, his letter was published anonymously since it challenged the popular philosophy of the day, Cartesian mind-body dualism, replacing it with empiricism and painting Diderot as an unfashionable secularist. These days, if you do that kind of thing, you just lose a few followers on *Twitter.*

We've developed many useful tools to characterize differences in neuroanatomy deep within the brain of a blind individual. In work carried out in UCLA's department of neurology, high-resolution brain imaging was used to characterize these structural differences in different groups of blind and sighted individuals. In work led by Natasha Leporé, a group that lost their sight under the age of five, a group that lost their sight over the age of fourteen, and sighted controls were imaged using MRI. The findings reported that typical areas associated with sight (occipital lobe) were decreased in volume, but many other areas of the brain were enlarged. In individuals who lost their eyesight when they were younger than five, the corpus callosum, which transmits visual information between the hemispheres of the brain, was much smaller compared to controls. This was considered to be the result of a still-developing myelin sheath in younger children. This myelin sheath creates a protective layer on the neural tracts, allowing them to fire more effectively; thus, the control group and the group who became blind after age fourteen had a fully developed myelin sheath and a larger corpus callosum.

But what about hearing? Is it possible to hone our attention on one specific conversation at a cocktail party? Neuroscientists and psychologists have framed our brain's ability to bring our auditory attention to focus on one stimulus in a noisy room. A listener has been able to process multiple streams of auditory information in order to identify the most relevant information for the individual. From work carried out by Dr. Monica Hawley at Boston University, we have learned that our auditory attention in the cocktail party problem works best as a binaural effect, requiring both ears. Individuals with only one functioning ear have more difficulty following multiple conversations and lack the ability to effectively situate the source of sound. This would suggest that binaurally processing and situating the source of stimuli may work together to allow you to hone in on it cognitively as well.

KILLMONGER VERSUS T'CHALLA: NATURE VERSUS NURTURE

WHEN: *Black Panther*

WHO: Black Panther, Erik Killmonger

SCIENCE CONCEPTS: Neuroepigenetics, gene–environment interactions

INTRODUCTION

There's a fatalistic idea that your genetics determine your health, suggesting everything depends on the sequence of chemical information coded in your genome. However, your genes are only part of the picture. In fact, you have control over many of the factors that make you healthy; there's more to it than just the genes you inherit from your mother and father. The sum of all your experiences and how they wire into your mental and physical health is partially a product of genes, but your environment comes into play too, and can affect your genes. Here, we will discuss some of the molecular mechanisms that shape the genome through experience and how they may have been the tragic fork in the road for Erik Killmonger and T'Challa.

BACKSTORY

In the events of *Black Panther*, two generations of Wakandans tread a tragic path of privilege and disenfranchisement. In Wakanda, a young T'Challa grows up with a loving royal family in a secret kingdom of wealth, technology, education, and resources. In America, a young Erik Stevens finishes a game of basketball in Oakland and is about to discover his father dead in his apartment. Both Erik and T'Challa come from a royal family but their respective fathers brought them into a world with completely different rules. Decades later both Erik and T'Challa have grown into incredibly intelligent and physically capable men. However, T'Challa becomes a king and defender of his nation while Erik takes the moniker Killmonger with the goal of bringing retribution to the injustices he witnessed growing up in America.

THE SCIENCE OF MARVEL

Erik Killmonger's childhood environment is a key motivating factor driving his actions in *Black Panther*. Despite being "the villain," he found vengeance and agency during his traumatic upbringing. In contrast to how Avengers deal with trauma and subsequent treatment, Killmonger faces adversity with resilience and the mind-set of what doesn't kill you makes you stronger. T'Challa, on the other hand, spent his childhood sheltered from violence, poverty, and social disenfranchisement. He was provided a safe haven where he had access to the best possible education befitting a royal family member in the most technologically advanced society on Earth. Given the environmental differences the two characters experienced, it's tempting to wonder what would have happened if T'Challa and Erik Stevens had been swapped at birth: would the events of *Black Panther* have transpired in exactly the same way? Here, we will focus on Erik Stevens (Killmonger) to explore the role environment plays in early childhood development, and we will also take a look at the molecular markers of traumatic memories and resilience.

After being orphaned, Killmonger had to survive in a low socioeconomic position in Oakland, California. Unlike today, in the 1980s violent crime, including a high murder rate, police malfeasance, and drug abuse were prevalent throughout the city. Stevens had a stressful upbringing. This does not only take a toll on a child's day-to-day existence; these stressors can follow the person into adulthood (in Stevens's case, to Wakanda). However, some would not have made it as far as Killmonger did. What made him the exception and can we trace it to changes in his brain? Similar to the symptoms of PTSD, there is altered sensitivity to stress hormones following trauma. Various genes and their products are known to change in response to social adversity and stress. Within the stress axis of the brain, the pea-sized pituitary gland at the base of the brain is responsible for synthesizing hormones that are released during trauma.

During trauma, corticotropin-releasing hormone is released from the hypothalamus-bound receptors on the pituitary, causing the secretion of adrenocorticotropic hormone (ACTH) and leading to increased sympathetic tone (fight or flight response). Corticotropin-releasing hormone (CRH) is a key component of the chained cascade of chemical messengers that communicate with the brain and body during traumatic experiences. We also know that genes, like CRH, have "on" and "off" switches for their function (see Hulk's Transformation). One example that we discuss in other entries is the way in which DNA can be marked with a chemical modification, methylation, that can effectively shut down production of a gene. Given Erik Stevens's motivations, it is clear that he refused to succumb to the environmental factors he grew up with. Unlike the way it can for many people, stress didn't incapacitate him or slow him down. Even though Stevens undoubtedly experienced grief and anger over the circumstances of his father's death, the chemical pathways that can incapacitate a person exposed to chronic stress had no effect on him. At the moment he found his father, N'Jobu, murdered he may have broken this neurochemical chain reaction by laying down DNA methylation onto CRH in the hypothalamus of his brain at a critical time during his childhood. This would partially explain how Stevens had the clarity and conviction to rise above the adversity he experienced in an environment like Oakland in the 1980s.

THE SCIENCE OF REAL LIFE

One of the seminal experiments that studied connections between early developmental environment and DNA methylation was conducted with rodents at McGill University. This work, led by Dr. Michael Meaney and Dr. Moshe Szyf, sought to understand whether DNA methylation plays a role in maternal care. While not entirely analogous to Erik Killmonger's journey, the experiment does offer insight into the molecular mechanisms at play during forms of social stress. In this study, mothers from one strain of lab rats would lick, groom, and nurse their young whereas the other strain did not. Following one week of being raised by each mother, pups would grow up to display different behaviors based on their first week of care. Animals that received multiple forms of care became less anxious than those that did not.

To assess whether this had anything to do with DNA methylation, the researchers sought out key regulators of the stress axis and measured the amounts of DNA methylation happening within the glucocorticoid receptor (GR). This receptor mediates stressful stimuli by relaying glucocorticoid release into electrical impulses between neurons. Too much of this receptor sensitizes an animal's behavior to small amounts of the stress hormone, whereas too little desensitizes their stress response. Animals that received no care had high levels of DNA methylation, and low expression of the glucocorticoid receptor, and vice versa. These results were even replicated in pups that were cross-fostered between mothers, showing that these effects are the result of maternal upbringing and not a separate hereditary effect.

The science discussed here is a simplification of a complex series of events that occur during social stressors such as being subject to a traumatic event, violence, or poverty. Our psychobiology and how it changes in response to such stimuli varies depending on our age, our experience, and our coping mechanisms. However, the field of epigenetics has offered some insight into how these processes generally function. Work carried out by Dr. Elliott Evans during his tenure at the Weizmann Institute of Science revealed the molecular changes that accompany social resilience in mouse models. In his experiments, he subjected mice to ten days of continuous social defeat in which a mouse would be introduced into the cage of an older, more aggressive mouse. These mice would then go on to display social avoidance behaviors due to being bullied. These bullied mice would then be placed in a new cage with a perforated divider; another mouse was kept on the other side of the divider. While most animals would avoid the interaction zone after being bullied, a subset would actually show a form of resilience. Animals that avoided their bully had less methylation in the CRH gene whereas animals that were resilient maintained higher levels of DNA methylation at this gene. DNA methylation mediated CRH gene expression, serving as a possible explanation for how typical social stress can lead to avoidance behaviors. While this is only a mouse, it does seem to model how certain aspects of resilience can be programmed into different tissues in the brain that elicit different behaviors. Nonetheless, even though DNA methylation has been shown to be an important mechanism underlying the etiology of mental illnesses, it is not an answer for everything. Psychosocial stress is a complex multifactorial issue; there are many pieces to this puzzle, ranging from genes and neurons to things we have yet to fully understand.

SPIDEY SENSE

WHEN: *Spider-Man: Homecoming, Captain America: Civil War, Avengers: Infinity War*

WHO: Spider-Man

SCIENCE CONCEPTS: Sensory neuroscience, spider cognition

INTRODUCTION

The sense of touch is an incredibly important one for humans since the neurons in our hands are mapped extensively onto our somatosensory cortex in our brain. We can sense pain, vibrations, mechanical movement, heat, and a variety of other tactile stimuli. How exactly is this process regulated, and is it possible to lower the threshold of sensation to perceive even more subtle changes from our immediate environment? How are these sensory signals transmitted in other animals such as spiders? Is it possible for these senses to alert us to danger in our own environment like Spider-Man's spider sense guides him?

BACKSTORY

Spider-Man has several spectacular powers such as super strength, wall-crawling abilities, and a unique precognitive "Spidey sense." This ability enables him to sense danger in the surrounding environment before it even happens. In *Captain America: Civil War* Spider-Man uses this ability to spot Ant-Man on Captain America's shield, to dodge an incoming Redwing, and to catch Bucky Barnes's punch. In *Spider-Man: Homecoming* he uses this sense to dodge a barrage of punches during the ATM robbery. During the events of *Avengers: Infinity War* Parker senses the moment when Thanos's Black Order lands on Earth several kilometers away.

THE SCIENCE OF MARVEL

To understand how Spider-Man can perceive his environment, we need to consider that Parker is still human but sensing his environment with the incredibly low threshold stimuli of a spider. For example, you may sense a breeze on the back of your hand, but if you were like Peter Parker you would sense that same breeze all over your body several seconds sooner. He's hyper-aware because his peripheral nervous system is hyper-innervated throughout his skin. As humans, we already have highly innervated hands, lips, and tongue, which we use for our tactile senses. These signal-ascending pathways pass through our spinal cord to the somatosensory cortex of our brain (see The Human Lie Detector). These senses involve several different kinds of neurons at the point where they receive a specific stimulus.

Imagine Bucky Barnes's closed fist flying toward Spider-Man. In that instant, before the punch was thrown, Spider-Man already had his hand up to catch it. Would it have been possible for Parker to have sensed that motion before it happened? Maybe. While that's not something a typical human could do, we do express certain proteins in our cells that may function as electrical sensors. For example, potassium channels such as Kir4.2 embedded in our cells can become activated with positively charged polyamines in our tissues and cells. A weak electric force (even one happening outside Parker's body) could be enough to polarize these polyamines and cause this channel to start transporting ions, causing an electric impulse. A cybernetic arm may be a big giveaway, but it doesn't rule out Parker's ability to sense any other muscle group winding up or contracting with great force. In either case, electrical activity is likely going to cause a contraction proportional to the force about to be exerted. So winding up any large forceful motion may be enough to set off Spider-Man's spider sense.

His sense isn't limited to electrical sensations. As we saw in *Avengers: Infinity War*, the hairs on his forearm allow him to sense danger in his surroundings. In this example, he may be experiencing low-threshold mechanoreception. Here, the hair follicles act like antennae for ambient vibrations that are innervated by hair plexus nerve endings. It makes sense that Peter Parker's hairs are extra innervated to detect subtle changes to his environment. Typically the threshold for sensation on a given hair would be relatively high, but for Parker, the slightest movement sends an electrical impulse through the nerve endings wrapped around the bottom of a given hair follicle to tell his brain that something is moving toward him. This sensation in conjunction with electrical sensing happening through polyamine-sensitive potassium channels in his skin helps modulate his reflexes so he's not constantly thinking he's in danger. For example, detecting a localized electrical impulse but no ambient movements toward him is like sensing a computer turning on nearby. Similarly, a breeze with no coincident surges in electrical activity in his immediate environment doesn't pose a threat. However, if he were to sense both of these things from directly behind him, it would urge his brain to initiate a sympathetic fight or flight response.

THE SCIENCE OF REAL LIFE

In real life we react the same way, but since in Parker's case this was due to a fateful spider bite, we can draw analogies to the manner in which most spiders sense their environment. For example, spiders have hair-like sensory organs called trichobothria, which are analogous to human hairs in their function except that each hair is innervated by its own nerve ending. This may align more with Peter Parker's spider-like qualities but it won't help him through the clothes he wears.

SPIDER WEBS

Interestingly, spiders are incredibly adept at sensing with more than just their body, through a decentralized cognition they make for themselves with their web. A spider that is active on its web will pluck at the threads it weaves onto its web to collect information about what parts of its web need mending or possibly to detect a trapped fly.

In a study at the Federal Institute of Bahia in Brazil, Dr. Hilton Japyassu sought to find out how effectively a spider can use its web to extend its sensory abilities. He looked at orb-weaver spiders, which make the typically shaped spiraling webs (see Spider-Man's Web-Shooters), and cobweb spiders, which have evolved a different type of hanging web for capturing prey.

In order to emulate this transition in behavior, Japyassu collected twelve different species of orb-weaver spiders in Brazil and clipped their webs to see if they would use their clipped orb webs as hanging webs for their prey. While not all the spiders adopted the strategies of cobweb spiders, some did. This raised the question as to where exactly the memory of this information is stored. Is it in the central nervous system or in a combination of the web and central nervous system? It appears that the web at least helps alleviate some of the cognitive load a spider uses to sense its prey. This makes you wonder why Spider-Man doesn't lay a large web over Manhattan's skyline in anticipation of a nearby siren, gunshot, or scream!

Interestingly, humans have a real spider sense, but it isn't exactly what you may think it is. In a study Dr. Joshua New of Barnard College asked participants to focus on three lines and to pick the longest one over three trials. After one round, participants returned to the same task only this time they were given a quick flash of an image for two hundred milliseconds. This image was of a hypodermic needle, a housefly, or a spider-like circle with radiating segments. Only 15 percent of the participants were able to notice the hypodermic needle, 10 percent the housefly, and more than half noticed the spider-like abstraction. Essentially the study suggested you are more likely to step on a hypodermic needle than a spider. This may come from our behavioral evolutionary roots in Africa, where a spider bite could have meant certain death.

HULK'S TRANSFORMATION

WHEN: The Incredible Hulk, The Avengers, Avengers: Age of Ultron, Thor: Ragnarok, Avengers: Infinity War

WHO: Hulk

SCIENCE CONCEPTS: Brain function, neuroscience, molecular biology

INTRODUCTION

Throughout human history we have developed complex social motivations for primal emotions such as happiness, sorrow, and anger. During our upbringing, we learn that there is often a time and place for these emotions and we should restrain them when they are detrimental to the social structures we uphold in our cultures (for example, it is not okay to cannibalize a coworker if he eats your lunch). These emotions can manifest themselves within areas of your brain, the neurons in these areas, and the genes expressed within these neurons. Here, we will discuss what these emotional changes look like and what we would expect to see happening in the brain of mild-mannered Bruce Banner as he transforms into the Incredible Hulk.

BACKSTORY

The most powerful (and angriest) Avenger is indisputably the Hulk. After undergoing a botched Super Soldier transformation using gamma rays (see Super Soldier Serum), Bruce Banner developed the ability to transform into the Incredible Hulk whenever he loses control of his emotions. As the Hulk, he is constantly enraged and driven to either "smash" or assert that he is the strongest one on the planet (see *Thor: Ragnarok*). This often leads to difficult situations where he poses a greater threat to the Avengers than to the villains he is fighting (see *The Avengers* and *Avengers: Age of Ultron*). To calm Hulk's nerves, fellow Avengers like Thor or Black Widow make attempts to settle him down with cognitive behavioral therapy (when that doesn't work, Iron Man can always subdue him under a collapsing skyscraper).

THE SCIENCE OF MARVEL

In the Marvel Cinematic Universe, Bruce Banner constantly struggles with his identity as the Hulk. Since Bruce Banner's emotional states are the catalyst for his transformations, we can probably trace how they begin by looking directly inside his brain. Moreover, these effects are likely to be very dramatic due to the contrast between a puny scientist who often avoids conflict and a brute with the power to knock out an Asgardian.

Considering gross brain anatomy, we will assume that changes in brain form relate directly to behavioral function. Thus we will expect various parts of the Hulk's brain to change size to accommodate less thinking and more aggression. In the Hulk, we are likely to see an attrition in the parts of the brain related to decision-making, higher-order cognitive processes, and executive function. Many of these processes are localized in the prefrontal cortex of the brain and allow impulse control, reasoning, and problem-solving. We may also see a loss of function in the cortical networks that process language since the Hulk often loses his ability to use prepositions and articulate how he's feeling. Other areas of the brain such as the amygdala and the limbic system may also shrink in volume; they process several basal emotions related to fear, anxiety, and anger.

As we focus on these regions of the brain, we also expect to see the rewiring of neuronal circuits that enforce aggressive behaviors and limit cognitive processes. These neurons function by connecting to each other through connections called synapses through which chemical and electrical information is relayed. When neurons fire together, they wire together and form circuits that process information from various sensory inputs (touch, smell, visual information, etc.) to initiate a specific behavior. For example, aggression is associated with hyperexcitability in the amygdala that may be reinforced by the increased release of neurotransmitters such as acetylcholine and glutamate. As these excitatory synapses fire in the amygdala, the synaptic strength connecting the neuronal populations in the prefrontal cortex to the amygdala would also weaken. In contrast, as Bruce Banner takes control over the Hulk, the synaptic strength of neurons connecting his prefrontal cortex to his amygdala and the limbic system provide the basis of a neural regression to Bruce Banner.

Lastly, we have to look directly within these changing neurons at the molecular level. Since the ability to become Bruce Banner or the Hulk is a reversible process, it stands to reason that its underlying molecular mechanisms are also plastic. In the hyperexcitable circuits projecting from the amygdala, we would expect an increased expression of the genes that produce excitatory neurotransmitters. These genes are subject to plastic regulatory mechanisms that shape gene function without changing the genes themselves and act like an on/off switch. Integrating these mechanisms within the cell and the neurons they connect to in various parts of the brain paints a picture of the Hulk's brain as well as some of the ways in which he can neurally control these transformations.

THE SCIENCE OF REAL LIFE

Thankfully, most humans exercise pretty good executive control over their aggressive and violent tendencies! In fact, much of this control is developed during childhood and adolescence within the prefrontal cortex. This part of the brain is developing until it reaches maturity in our early twenties and we often attribute this work in progress to some of the risk-taking behaviors teenagers have. Outside of normal brain development, exposure to toxic chemicals such as tetrachloroethylene can lead to hyperactivity and aggressive behaviors in children. Similarly, social factors related to childhood abuse can steer brain function, its anatomy, and its cellular functions toward aggressive behaviors.

A CHANGE IN PERSONALITY

In non-superhero adults, we can sometimes see dramatic changes in personality and a loss of executive control in certain very unusual situations. One such case was that of Phineas Gage (1823–1860). Gage survived an injury in which a metal pipe went through his skull and destroyed a large amount of his left frontal lobe. Against all odds, he recovered but became completely different in his personality. His attending physician, J.M. Harlow, described him as "fitful, irreverent, [and] indulging at times in the grossest profanity." This is one of the earliest pieces of evidence describing the function of the frontal cortex in regulating executive function and how it can affect personality.

Neuroscientists have adopted different animal models to understand how behavior is linked to certain areas of the brain. The African cichlid fish *Astatotilapia burtoni* is capable of its own version of "Hulking out." While not nearly as dangerous as the Hulk (but maybe as aggressive), males of this species exist in one of two social states: a Hulk-like dominant or a Bruce Banneresque subordinate. Dominant males are brightly colored, with dark stripes underneath their eyes and along their throats and pectoral fins. These males bite, chase, and threaten tankmates in order to secure territory and mates. On the other hand, the subordinate male is very drably colored and spends most of his time hiding and avoiding the dominant male. Now, if you take a larger subordinate male and put him next to a smaller dominant male, their behaviors switch. The subordinate male will immediately start showing aggressive behaviors, change his colors, and turn on his eyebar and dark colorings (way more impressive than just turning green). Over hours, their internal physiology changes as the ascending male produces more testosterone analogs and the descending male experiences stress-induced cortisol peaks. In their skulls, neurons of the ascending males rewire, size up, and fire up circuits that are involved in aggressive and reproductive behaviors.

Chapter 4

Fantastic Physiology

SUPER SOLDIER SERUM

WHEN: *Captain America: The First Avenger, The Incredible Hulk, Captain America: The Winter Soldier*

WHO: Captain America

SCIENCE CONCEPTS: Genetics, gene editing, pharmacology, nanocarriers

INTRODUCTION

How can someone achieve peak human physique? Olympic athletes require rigorous training regimens and years of dedication to arrive at a state that hones a single skill to perfection. During this training, the human body tunes its physiology, its cells, and the expression of different genes in different tissues to perform a variety of functions. If muscle growth is required, a series of growth factors will cause the proliferation and the growth of new cells and tissues. If stamina needs to be increased, oxygen-carrying loads in red blood cells will rise, making them more effective for cellular respiration. If increased agility and reflexes are required, the brain will form new synapses between neurons in relevant motor cortices to provide the cognitive proficiency to do backflips on demand. However, in the case of Captain America we can find shortcuts to improve all of these skills at once through targeted gene therapy and controlled activation of the mysterious Super Soldier Serum.

BACKSTORY

In *Captain America: The First Avenger* we see a frail Steve Rogers achieve peak human physique as the mighty Captain America in a procedure that lasts about five minutes. In order to understand Steve Rogers's transformation, we need to understand how the scientist behind the Super Soldier Serum, Abraham Erskine, formulated, delivered, and activated this process. Aside from a not-so-stringent selection process, it involved administering the serum orally, intravenously, and a final activation through the exposure to Vita-Rays. Following this treatment, Steve Rogers immediately gains muscle mass, develops super speed and agility, and even increases in height. So what goes into a Super Soldier Serum? Why does it require so many routes into the body, and why must light be used to activate it?

THE SCIENCE OF MARVEL

When done right, Super Soldiers in the MCU undergo a transformation that makes them the best humans they can be (e.g., Captain America). They increase their muscle mass, speed, agility, and even the molecular density of their skin and muscle tissues through synthetic proteins they acquire from the serum. In order to hack human biology so deeply we would need to focus on how to manipulate genes within a given cell. These genes make up your genome, which is nigh identical in every cell in your body. This genome is an important blueprint that maps out the many functions that cells in your body can have. Your genome is made up of deoxyribonucleic acid (DNA), which codes for a ribonucleic acid (RNA) message that ultimately gets turned into a functional protein. Simply put, your genes (DNA) are capable of sending messages into your cells (via RNA) to achieve a certain function (via proteins). Every cell of your body uses your genome differently to produce the right messages, to make the right proteins, and to make a given cell function the way it should. In Abraham Erskine's Super Soldier Serum, there must be some form of tissue-specific control of gene function that allows a human to achieve peak physique, while introducing new genes that push the extremes of superhuman ability.

When it comes to muscle enhancement we can imagine a targeted knockout of the genes that encode myostatin, insulin growth factor (IGF), or alpha-actinin-3 (ACTN3). These genes produce important proteins, which are key signaling molecule pathways that control muscle growth. Each of these genes has well-known and localized expression in muscle in healthy tissues, but playing with their functions can have extreme effects on muscle metabolism. Mice that have been genetically engineered to have a defective myostatin gene develop muscle mass at a significantly accelerated rate. Naturally occurring mutations in the IGF gene in humans and certain breeds of bull also increase muscle mass far beyond what is normally expected for a healthy adult. Interestingly, known mutations in the ACTN3 gene in humans have been suggested as a cause for variation in athletic performance and sprinting abilities.

When it comes to increasing stamina and endurance, it would be reasonable to assume that Captain America's body is particularly efficient at processing oxygen for cellular respiration. These changes would probably go hand in hand with peaks in erythropoietin (EPO) expression, which is a key regulator in blood cell proliferation. EPO is known to increase the production of red blood cells, resulting in an increased oxygen-carrying load. Oxygen-carrying load is often measured by hematocrit percentage, which is the volume of red blood cells in your blood. In a healthy adult, this value is 34–50 percent, but use of EPO for "blood doping" can drive hematocrit as high as 80 percent.

GENES DO THE WORK

Note that we name only a handful of genes of interest that regulate some of these processes related to muscle mass and strength when in fact many more proteins regulate these tissue functions. A tighter control over these processes would require regulating several hundred types of genes in various cells and tissues. Additionally, we would need to introduce new genes that would encode synthetic proteins. Achieving this objective necessitates the presence of something else in the Super Soldier Serum that allows new genes to be stably introduced into different cells. Vita-Rays play this role during the controlled administration of the serum, enabling the targeted regulation of tissue-specific activity.

THE SCIENCE OF REAL LIFE

At one point, you stood in a classroom with your peers and recognized a range of abilities. Some of you could jump the highest, run the fastest, or play the best game of chess. The natural variation in these traits is a mix between what you are born with and how you interact with your environment. How can we resolve these traits (and their extremes) to something that can be engineered and administered to make a Super Soldier? Within the approximately twenty thousand genes that humans use to code for protein, we have been successful in identifying a few of the important ones that could contribute to different "super" traits. However, this research often describes how a *single* gene in a mouse can influence a given trait (and we are not mice). Even then, the technology used to manipulate genes is very invasive and can involve removing embryos from the womb or injecting a large needle into a given tissue. These approaches also come with the challenges of targeting the right gene in the right part of your genome. In the case of Captain America, you would need to modify many different genes in different parts of the body.

With these caveats in mind, there have been attempts to carry out gene therapy to modify the human genome and cure disease. However, deaths due to these treatments have slowed progress and garnered scrutiny in the field as a whole. In one of these cases, a patient developed leukemia, suggesting that the modification of genetic material caused mutations in the wrong tissues. In another case, treatment resulted in a fatal immune response, implying that the way the treatment was delivered, through a virus, may be dangerous. Despite this, new technologies have shown promise.

Considering the hurdle of nonspecific insertion of genetic information that may result in leukemia, the recent development of clustered regularly interspaced short palindromic repeats (CRISPR) genome editing has revolutionized how we target the genome with surgical precision. This system is adapted from bacteria, where it is used to mutate unwelcome viral DNA insertions. CRISPR functions to recognize and bind a guide template and then proceeds to systematically read your genome until it finds a match to its guide template. It then acts like a pair of molecular scissors and starts cutting DNA in order to inactivate a gene and, in some cases, introduce new ones. It's a lot like your kindergarten arts and crafts table, except at a molecular level. While this technology is about a decade old, it has developed quickly and has led to the creation of several biotech companies (Editas Medicine and Caribou Biosciences, for example) that are promising a new age of therapeutics. In 2018, the first gene-edited babies were made in China by Dr. He Jiankui to increase resistance to HIV, demonstrating this technology is actually feasible in humans. In basic research labs, CRISPR has also proven to be a relatively simple technique compared to the other approaches we've been using for the last thirty or more years. Whereas some tools were only effective in mice, CRISPR has been successful in all kinds of animals (mammals, fish, insects, etc.) and can be used to target several genes at once.

Assuming you can target the genome with some precision, it remains difficult to make sure you can target the right tissue while also avoiding any unwanted immune responses (as was the case in initial gene therapy trials). It is no simple task to deliver a protein like the one used in CRISPR to any given tissue, and it is of critical importance to deliver the right modifications to the right tissues. As such, modifying genes for strength in the brain would serve no purpose and can indeed damage basic brain function. Although tissue- and cell-targeted therapy for drug delivery was first conceptualized back in the early 1900s, the study of molecular delivery systems for this purpose did not begin until the late 1970s. Due to advances in nanotechnology, the development of nano-sized drug carriers, known as nanocarriers, has escalated dramatically over the last decade.

Nanocarriers represent a new class of synthetic particles that can be used to deliver drugs to precise targets within the body. They are highly versatile delivery platforms, ranging from 10 to 1,000 nanometers in size, and show promise in reducing undesirable immune responses when compared to other potential drug-delivery routes, such as viruses. To maximize drug potency and minimize potential off-target side effects, it is advantageous to trigger cargo release at the site of interest. For example, little fat bubbles (called liposomes) are produced naturally in your cells and are used to move cellular cargo without eliciting an immune response. These liposomes can be generated synthetically as nanocarriers to store CRISPR components and look like normal cellular machinery in a given tissue. Certain modifications to liposomes can also sensitize them to light at different wavelengths, causing them to disintegrate under the guiding hand of localized light (or a laser). This technology (termed photodynamic therapy) is an application that's very similar to Abraham Erskine's Vita-Rays; it opens therapeutic venues that allow tissue-specific activation and reduced immune reactions.

CRYOSTASIS

WHEN: *Captain America: The First Avenger, Captain America: The Winter Soldier, Captain America: Civil War*

WHO: Captain America, Winter Soldier

SCIENCE CONCEPTS: Cryobiology, hibernation

INTRODUCTION

The animal kingdom has developed an array of strategies to adapt to extreme environments. In the freezing cold, various species can lower their body temperature, regulate their metabolism, and physiologically turn off for an entire season. For example, the Alaskan wood frog completely freezes in the winter and emerges from cryostasis in the spring season. Small mammals have also learned to depress their metabolism in order to wait out the cold and harsh winters by burning fat they stored up in the late summer. While these strategies may work for frogs and squirrels, could we use the cold to pause and extend human life? Implementing these adaptations to organ banks could save lives. Will it ever be possible for humans to be like Captain America and the Winter Soldier, living across decades without aging a day?

BACKSTORY

In the final standoff between Captain America and the Red Skull, Steve Rogers chooses to crash land a HYDRA bomber, *Valkyrie*, into the Arctic Ocean. Upon impact he is thrown into the Atlantic and lost for approximately seventy years. During this time, Steve Rogers remains frozen in ice until S.H.I.E.L.D. discovers him still alive and asleep in cryostasis. Similarly, Bucky Barnes is also subjected to cryostasis as he becomes the Winter Soldier. After becoming a Super Soldier himself, Bucky is repeatedly freeze-thawed across time in order to carry out assassinations throughout history. So what happens to the human body when its temperature drops, and what is allowing Steve Rogers and his BFF Bucky to resuscitate from the cold?

THE SCIENCE OF MARVEL

Bucky Barnes and Steve Rogers undergo variations of cryostasis, allowing them to survive for decades without aging. In Steve Rogers's case, his crash landing into the Arctic Ocean resulted in a seventy-year-long nap that was interrupted by S.H.I.E.L.D. agents. In contrast, the Winter Soldier is kept in a containment pod where his cooling is finely controlled for several rounds of torpor and arousal throughout human history. Unlike these Super Soldiers, we would die from hypothermia if our core body temperature dropped and stayed below 98.6°F (37°C). Our organs would shut down, leading to respiratory and heart failure. How did these Super Soldiers survive the cold? What happens to us when we freeze, and how do some animals protect themselves against the harsh cold?

If I were to cool you down without protective clothing you would succumb to hypothermia. Over time you would suffer a loss of speech, drowsiness, cognitive decline, and general clumsiness due to muscle stiffness. At your extremities, you would suffer from frostbite, which would successively become red, pale, numb, and blue. The most damaging effects of freezing happen with the formation of ice in your tissues. This ice causes damage through the shearing force of crystal formation and the dehydration of cells. The flow of water out of the cell to form ice would also result in an increase of intercellular solutes, which would offset the homeostatic balance of cellular chemistry. These effects are irreversible and are the main hurdles that Captain America and the Winter Soldier need to overcome.

In nature, many animals can survive dramatic drops in temperature through a variety of physiological changes. For example, wood frogs minimize the formation of ice within their cells by signaling glucose production in the liver and pumping it into the cells of their tissues. The influx of intercellular sugar traps water in cells and restricts ice formation between cells to the water that already occupies that space. This sugary antifreeze reduces cellular dehydration and allows a frog to stay alive in freezing water. This physiological strategy could have made sense if Captain America's super liver was capable of storing enough glucagon (stored branched glucose) that could last seventy years of sleep (anyone who knows him personally knows Steve Rogers is a pretty sweet guy). Similar strategies involve using water-loving antifreeze proteins (such as those found in arctic fish) that bind water before it forms ice microcrystals in the animal's blood. If Erskine planned to have Super Soldiers put on ice, it would have made sense to include these genetic plans in the serum injected into Steve Rogers (see Super Soldier Serum).

In addition to protection from ice formation, other mammals have learned to thermoregulate and turn their metabolism on and off through bouts of hibernation. For example, small mammals like the thirteen-lined ground squirrel reduce their metabolism by limiting their respiration and reducing their heart rate from 200 bpm to just over 5 bpm. These changes result in a low energy economy where the body temperature fluctuates between normal body temperature and a body temperature that matches the ambient environment (<32°F [0°C]). In order to maintain these fluctuations, Steve Rogers would have to have fat stores or food caches to sustain him for seven decades (sneaking in rations of HYDRA chocolates stored on the *Valkyrie*). Interestingly, most mammals undergo these physiological changes as they are slowly cooled, but suffer the stresses of ice formation in their tissues. If a Super Soldier was programmed to express cryoprotectant antifreeze (whether it be as a protein or sugar catabolism), his body would enter a dormant state that would match that of a hibernating animal.

THE SCIENCE OF REAL LIFE

The fantasy of cryostasis that Bucky Barnes experienced is something that many humans would love to turn into a reality; various approaches to using temperature and cryoprotectants to preserve tissues over long periods of time have been explored. In terms of real-life "hibernating" by a human, there's the curious story of a fourteen-year-old Swedish girl, Karolina Olsson, who purportedly slept for thirty-two years, 1876–1908.

GET YOUR BODY FROZEN

A handful of companies offer the opportunity to cryopreserve your body today in the hope that you can be resuscitated in a future where medical technology has advanced to a level where any illness can be cured. The Alcor Life Extension Foundation has preserved 158 human bodies since 1966 (with a waiting list of 1,194 subjects waiting to be preserved). Their first patient was James Bedford (1893–1967), a University of California psychology professor who was diagnosed with an aggressive kidney cancer that had metastasized to his lungs. Following his legal death, he was cryopreserved with dimethyl sulfoxide and is currently cooled below 32°F (0°C) in Scottsdale, Arizona.

CHAPTER 4: FANTASTIC PHYSIOLOGY

While it still remains difficult to ascertain if a body can be resuscitated after it is cooled below 32°F (0°C), some progress has been made on a smaller scale. For example, we are capable of cryopreserving small tissues and populations of cells for long-term storage well below −112°F (−80°C) (e.g., sperm, egg cells, and human embryos). These approaches employ the use of vitrification, where cryoprotectants inhibit the formation of ice and slow the movement of molecules in a sample, keeping it in a solid/liquid state at temperatures below −284°F (−140°C). Scaling these approaches to whole tissues and organs still remains a great challenge; this is an area where more rigorous research is needed. For example, most organ banks can preserve tissues for four to twelve hours (and never below 32°F [0°C]). If these organs can be safely preserved at subzero temperatures it would allow many more lives to be saved by transplants and a more effective distribution of tissues over time and place. Unfortunately, many of these cryoprotectants are toxic in the concentrations needed to preserve larger tissues and would require a level of additional detoxification that is not possible with today's medical technology.

Currently, many different groups take part in the preclinical side of research by developing a better understanding of model systems (e.g., those of the thirteen-lined ground squirrel, arctic fish, or woodland frogs) that can initiate hibernation. For example, I have measured how a molecular mark called DNA methylation can program the expression of genes related to muscle metabolism during bouts of arousal in hibernating ground squirrels. DNA methylation acts like an off switch and can turn off a gene important for muscle metabolism (myocyte enhancer factor 2C) during torpor. While these processes probably involve many different layers of regulation, a basic understanding of some of these mechanisms would allow us to develop drugs to target specific human tissues and stimulate a controlled metabolic depression. For example, cataloging the genes that are expressed in different tissues of a hibernator that are shared with the human genome could provide us insight into the molecular changes that grant a hibernator the ability to survive extreme cold.

PHARMACOLOGY OF THE HEART-SHAPED HERB

WHEN: *Black Panther*	
WHO: Black Panther, Erik Killmonger	
SCIENCE CONCEPTS: Plant physiology, plant evolution, toxicology	

INTRODUCTION

We have coevolved with several plant species to create crops that serve many human needs. For example, we have selectively bred strains of maize to make cultivars of corn for human consumption or we can use the large generated biomass of corn husks to create biofuel. One of the oldest applications of different plant species stems from our use of specific plants for medicinal use. In fact, many of today's most-used drugs were derived from a plant source then scaled up through chemical synthesis to serve larger populations (for example, opioids, digoxin, and aspirin). To this day, we are still screening compounds made from different plant species to identify new drugs that can be used to remedy the symptoms or causes of different diseases. Here, we will discuss how the Heart-Shaped Herb may have evolved in Wakanda and how its compounds imbued T'Challa with the powers of the Black Panther.

BACKSTORY

In *Black Panther*, each incumbent ruler of Wakanda ingests the Heart-Shaped Herb to grant them superhuman powers to protect their country. This naturally sourced non-GMO plant has a wide array of effects on its user, similar to what the Super Soldier Serum does for Steve Rogers's biology. This is evident in *Avengers: Infinity War*, where we see T'Challa and Steve Rogers match each other's superhuman speeds as they attack Thanos's army of outriders. In his transformation to Black Panther, T'Challa is granted enhanced agility, endurance, strength, cognitive acuity, and a hallucinatory visit to his ancestors. The way in which this plant imbues powers is a bit of a mystery, but it relies on the transitive powers of vibranium biosequestered by the herb in the groves on Mount Bashenga. Just as these powers can be given, they can also be taken away with another herb preparation used during ceremonial combat for the title of Black Panther.

THE SCIENCE OF MARVEL

The discovery of the Heart-Shaped Herb by Bashenga led to its use to create a lineage of Black Panthers that would protect Wakanda from outsiders. How might have such a plant evolved? Most plants produce macromolecules that they need for their primary metabolism for growth and development. This includes the synthesis of amino acids to make protein, carbohydrates to store energy, nucleic acids to code genetic information as well as several fats and oils. In addition to these primary metabolites, some families of plants also produce secondary metabolites that are not required for growth and development but provide a chemical line of defense that increases its fitness in a niche environment. Given its cave-like environment, we can speculate that the Heart-Shaped Herb is producing secondary metabolites that may inhibit the growth of other plants (mosses, ferns, liverworts, etc.) or toxins that can affect herbivore food choices. Many species of plants that produce neurotoxins to deter herbivores concentrate their toxins in leaves or reproductive organs (pistils, stamens, etc.), carrying a higher penalty for their ingestion. In the Heart-Shaped Herb, it would make sense that the cocktail of secondary metabolites that facilitate a Black Panther transformation would be found in the syrupy nectar in its purple flowers.

These secondary metabolites can be derived from a variety of biochemical pathways and likely target several parts of T'Challa's physiology when he consumes the herb. Depending on the biochemical pathway being used to make these metabolites, they can be classified as flavonoids, terpenoids, nitrogen-containing alkaloids, or sulfur-containing compounds. For example, one secondary metabolite may act as a drug for a receptor that modulates stamina in his lungs, whereas another metabolite may act as a drug for a receptor that improves cognitive acuity in his brain. When we consider typical receptor biology in T'Challa, we can imagine one of these secondary metabolites acting like a drug with high specificity to different receptors in his tissues. For example, these metabolites may mimic some of the chemistry seen for adrenaline, allowing him to increase his sympathetic tone and sustain an adrenaline rush. Similarly, other secondary metabolites can share the chemical structure of endogenous molecules that regulate nociception or perceptions of acute pain, increasing his pain thresholds in battle.

One particular effect of such secondary metabolites may explain some of the immediate effects of the herb when ingested by T'Challa and Erik Killmonger. During their ceremonial transformation, both were buried in red clay and visited their subjective versions of an Ancestral Plane, where each had a conversation with his deceased father. While possibly a mystical property of the herb or the vibranium therein, this most closely resembles a hallucinatory episode. Several secondary metabolites derived from the chemical family of alkaloids (the largest groups of secondary metabolites) can elicit hallucinations by binding serotonin receptors in the brain, signaling hyperactivity in the prefrontal cortex. This could offer some explanation for the similarity of Killmonger's and T'Challa's experience despite the vast differences in their backgrounds. Several species of plants have evolved such neurotoxins independently over the course of some six hundred million years to deter herbivores.

CHAPTER 4: FANTASTIC PHYSIOLOGY

THE SCIENCE OF REAL LIFE

Humans have relied on medicinal herbs since our earliest history. Ancient Egyptians used willow bark to treat aches and pains and Hippocrates even documented its use to fight off fevers. Willow bark's active compound, salicylic acid, was eventually discovered and purified by French pharmacist Henri Leroux in 1829. In 1874, Hermann Kolbe figured out how to chemically synthesize salicylic acid, which was further modified by chemist Felix Hoffmann at Bayer. Salicylic acid began to be distributed to doctors in 1899 and became an over-the-counter drug, aspirin, in 1915. The cultural origins of the use of willow bark may be older even than our recorded history, but it provided us with a drug that has been widely used around the world for the last century.

Upon examining the altered physiology of T'Challa we can point to several plants that when combined could potentially contribute to T'Challa's altered physiology. For example, changes in sympathetic tone can be elicited through the ingestion of the Chinese desert shrub *Ephedra equisetina* (ma huang), which produces secondary metabolites ephedrine and pseudoephedrine. Ephedrine acts on the same receptors as epinephrine, making it an effective stimulant. This herbal extract has been banned by several antidoping agencies because of its ability to enhance an athlete's performance. In controlled amounts plant extracts of ephedra have been used in traditional Chinese medicine for centuries to treat bronchitis and asthma. Similarly, pain thresholds can be lowered by opioids extracted from poppies but the treatment comes with a heavy burden: possible addiction and substance abuse. Hallucinatory effects of several plants are also common in spiritual and religious ceremonies that involve the ingestion of peyote or San Pedro cacti. These cacti contain mescaline, a drug that is known to bind serotonin receptors in the brain with a high affinity, causing hallucinatory episodes. In most of these cases, however, the effects of an herb or its active ingredients last only as long as the substance stays in the user's system. Unlike the Heart-Shaped Herb, with pretty much permanent effects, several of these plants need to be ingested regularly to maintain their effects.

We've discovered many drugs looking at plants for chemical inspiration and we have expanded our survey into more exotic model systems in attempts to find new drugs. For example, the venoms of many snakes and marine invertebrates have evolved molecular strategies to depress the central nervous system and speed up the cardiovascular system. In a predator-prey scenario, this allows the venom to accelerate its spread while often paralyzing the prey. While deadly in the venom glands of a pit viper or cone snail, venom has been used by pharmacologists in the laboratory to synthesize beneficial treatments for humans. For example, in 1975, venom separated from the Brazilian pit viper was shown to act as an anticoagulant and led to the discovery and development of captopril, one of the first oral treatments for hypertension. Needless to say, a Heart-Shaped Herb that has a range of beneficial effects without any adverse side effects is as unusual as a serum that can create a Super Soldier.

EXTREMIS AND TISSUE REGENERATION

WHEN: *Iron Man 3, Agents of S.H.I.E.L.D.*

WHO: Aldrich Killian, Pepper Potts, Eric Savin, Ellen Brandt, Jack Taggart, Chad Davis, David Samuels, Deathlok, Scorch, Brian Hayward, Quake

SCIENCE CONCEPTS: Tissue regeneration, stem cells, autonomic brain function

INTRODUCTION

We are able to regenerate many different tissues in our body. For example, we replace about ten pints of our blood over the course of a few months, and a human liver can be split from a donor and regenerate to its original size and function in about two years. However, there are also parts of our body that do not come back if they are damaged or removed, such as whole limbs and brain tissue. In nature there are many animals capable of recovering whole limbs, and others capable of rapid tissue regeneration.

BACKSTORY

In *Iron Man 3* and *Agents of S.H.I.E.L.D.*, the Extremis serum grants its users superhuman abilities such as increased strength, faster reflexes, and the ability to generate extreme heat. However, it was initially designed as a way to modify the human brain to allow a new level of control over biological processes throughout the body. Extremis was initially tested on disabled and injured veterans, where it was able to grant one of its users the ability to grow a new arm within a few minutes. During the growth of a limb the user's metabolic heat increases dramatically, resembling a burning ember as the tissue grows to its full size.

THE SCIENCE OF MARVEL

In *Iron Man 3*, the Extremis serum is originally engineered by Dr. Maya Hansen as a way to unlock the ultimate potential of a living organism. Despite its promise as a therapeutic, the serum suffers from stability issues when used in plants, leading to spontaneous combustion. This leads Hansen to seek help from Tony Stark to stabilize its effects. Eventually, the technology is developed under the funding and support of Advanced Idea Mechanics (AIM) and the supervision of Aldrich Killian. At AIM, the Extremis serum is wired within unused parts of the brain to provide direct control over autonomic biological functions. The use of Extremis-enabled abilities causes increased exothermic properties; this leads to different parts of the body becoming extremely exothermic, often burning white hot.

One of the most important uses of the Extremis serum is its direct integration into the brain's control over the body. This procedure corresponds to a physical location in the brain or "open slot" that can be used to facilitate the serum's powers. However, there is a tremendous amount of variability in a host's ability to accept treatment. This variation may be due to the brain state of a subject. In people such as Ellen Brandt, an amputation has long-lasting effects on her brain and how her arm has been represented in the primary motor cortex. Typically this neural representation extinguishes over time, but there is a window of time in which it can still be repurposed for other uses. However, the introduction of Extremis and its timing may be the key to its successful integration in these motor cortices. In a similar fashion, Aldrich Killian suffered from several debilitating handicaps throughout his life, possibly resulting in the stunted development of different parts of his brain that can be "repurposed" by Extremis.

Once integrated into the brain of a host, Extremis is able to wire conscious thought into the basal autonomic function of the person's body. These processes are regulated by the autonomic nervous system, which projects nerves into all of our major organs and tissues. These processes oversee things that don't usually require conscious thought, such as breathing and digestion. Extremis facilitates this process by further allowing the user to recode and rewrite the chemistry of the DNA within the cells of these tissues. What changes to our DNA and the expression of our genes allow for tissue regeneration? In order to consciously regrow bone, muscle, and connective tissue, we would need to emulate the regenerative processes seen in axolotl salamanders. When a limb is removed from one of these salamanders, they dedifferentiate the cells near the site of injury to create a blastema. This involves the expression of genes that act as master regulators of many other genes in the nucleus of a given cell (specifically the transcription factors Oct4, Nanog, and Sox6). This process causes differentiated cells with a specific function to regress to a more stem-like state in which they have the potential to assume many different functions; we call this an induced state of pluripotency. Pluripotent cells within the blastema proceed to divide and grow in a similar fashion as they did during embryological development.

THE SCIENCE OF REAL LIFE

Unlike Extremis users, our brains do not have executive control over something like tissue regeneration. We need an abstraction of these systems in order to avoid muddling our own cognitive processes. Think about it this way: you don't need to perceive the circuitry and organization of light passing through every pixel on an LED screen in order to use a smartphone. Our autonomic nervous system is divided into two parts, the parasympathetic (resting state functions) and sympathetic (fight or flight functions) systems. One way our brains have an ability to direct healing is through the vagal nerve (a part of our parasympathetic system) when it wires into our spleen. The spleen is responsible for recycling red blood cells and storing and releasing white blood cells needed for an immune response. Vagal stimulation is capable of regulating the release of these white blood cells when fighting off an infection. This process can also initiate an inflammatory response that may give you a fever, but you won't spontaneously combust (thankfully).

When we regenerate tissue we have a cascade of complex molecular, cellular, and physiological processes that may allow small populations of tissues to regenerate. Much of this depends on the available pluripotent cells that we have within our body. For example, children and infants would be more likely to be capable of regenerating whole digits compared to adults due to the availability of these reserve stem cell pools in the young. These cells are quite capable of restoring tissues and organs during embryogenesis and fetal development but are not as effective once we are older. One strategy currently being developed is to build new organs for replantation using ex vivo technologies. For example, we can take fat or skin cells from a patient, culture them in the lab (in vitro), and expose them to factors that can dedifferentiate them into a pluripotent state. This is an analogous way of generating our own blastema in a petri dish under the control of lab conditions and can be used to fill in gaps created by injuries in different kinds of tissues. Some labs are even going as far as combining cultured cells and 3D-printed scaffolds made from structural proteins or sugars to create synthetic organs. While these 3D-printed organs have not yet been used clinically, they have been successfully used to make functioning kidneys for rat model systems in the lab.

While promising, current research related to generating pluripotent stem cells uses very laborious and time-consuming processes, often yielding results slowly. It requires reprogramming adult cells outside of the patient with viruses and transfection reagents. One novel alternative is to switch the function of cells in tissue (without regressing them to a stem-like state) through the process of transdifferentiation. One approach of this concept can be carried out with tissue nanotransfection, which uses a topically applied array of nanotubes (a nanochip) and factors that can reprogram a cell (genes of interest). The nanochip generates a localized electric pulse that creates temporary nanopores in the cell membrane that allows the uptake of factors that can transdifferentiate the cells. This approach has been used in mice to reprogram cells in the skin to become endothelial cells that then grow into the capillary vasculature through the increased expression of the genes Etv2, FoxC2, and Fli1. In fact, when mice received a surgery that effectively killed their leg by cutting off its access to blood, tissue nanotransfection was able to regenerate the damage done by the surgery. While this type of technology may be ten to twenty years from the clinic, it offers fantastic innovations that could transform regenerative therapy.

CHAPTER 4: FANTASTIC PHYSIOLOGY

IMMORTALITY

WHEN: *Captain America: The First Avenger, Captain America: The Winter Soldier, Captain America: Civil War, Thor, Thor: The Dark World, Thor: Ragnarok, Avengers: Infinity War*

WHO: Asgardians (Thor, Odin, Loki, Heimdall, Hela, etc.), Titans (Thanos), The Ancient One

SCIENCE CONCEPTS: Longevity, molecular biology, age-related pathophysiology

INTRODUCTION

One hundred years ago, the average life expectancy for a man in the United States was fifty-three-and-a-half years and for a woman was fifty-six years. We expect US life expectancy by 2050 to be 89–94 for women and 83–86 for men. Variables such as gender, genetics, access to healthcare, diet, and exercise all contribute to rising life expectancy. While advances in these factors improve the overall quality of life, the exact way in which these changes manifest themselves at a molecular level in our cells has only been recently revealed to molecular biologists. At an organismal level we've learned quite a bit about longevity by studying animals that live very long or very short lifespans. Equipped with this information, we can speculate how Asgardians and a Sorcerer Supreme can live for centuries.

BACKSTORY

Throughout the Marvel Cinematic Universe different individuals are considered timeless when compared to the typical lifespan of a human being. For example, in *Doctor Strange*, The Ancient One draws her power from the Dark Dimension in order to extend her life well beyond seven hundred years, while still maintaining the vitality and resilience of a seemingly middle-aged human. Individual Asgardians have lived since the thirteenth century, more than eight hundred years ago. Despite their many years alive, Asgardians such as Loki and Thor are still considered to be juvenile relative to an even older generation of Asgardians such as Odin, Hela, and Heimdall. These characters have several unique abilities, and through the ages they have learned to wield their powers while adding to their knowledge and enhancing their skills.

THE SCIENCE OF MARVEL

Many characters in the Marvel Cinematic Universe age beyond the years that the oldest humans achieve in a lifetime. At a cellular level, aging is often understood through the phenomenon of senescence, or a cell's inability to divide over time. A typical human being divides its cells 10^{16} times over the course of a lifetime. This considers the rates of replication that differentiate cells into their specified tissues and the regenerating pools of stem cells that typically maintain our health and growth. Is it possible to delay the onset of cellular senescence? In thinking about the answer to this question, we have to consider that over ten quadrillion divisions the "quality" of our DNA degrades in a variety of ways. For example, the genome in a given cell can collect mutations or inefficiently correct these mutations when they occur.

Let's say that, as in humans, within the nucleus of a given cell in a healthy young Asgardian like Thor there are twenty-three pairs of chromosomes that store the hereditary information he received from his parents. These chromosomes have a conserved architecture that maintains their stability and function. When you consider the X-shape of most human chromosomes, the belt of the X is called the centromere while the four capped tips of the X are called the telomeres. When DNA replicates through the process of mitosis, this material doubles with each successive round of cell division and over time telomeres shorten. Eventually the telomere cap shortens to a point at which the cell is unable to divide, contributing to senescence. To attain such extended lifespans, the many long-lived characters in the MCU need to overexpress a highly efficient enzyme known as telomerase. Telomerase replenishes the caps of chromosomes by adding nucleotides bases (or the "rungs" in the DNA double helix) to telomeres to stop them from shortening.

Over time, DNA itself is subject to oxidative damage done by free radicals, resulting in the accumulation of mutations. Of all the tissues, the aging brain is subject to high amounts of oxidative damage due to its higher metabolic rate, increased consumption of oxygen, and lower antioxidant capacity. Considering the wise and all-knowing Ancient One, we can speculate that her ability to retain her wisdom may at the very least involve countering the cumulative oxidative stress collected during more than seven hundred years of living. This can be achieved by the expression of known antioxidant proteins such as superoxide dismutase, catalase, or glutathione peroxidase. Even when this damage occurs, a family of highly efficient DNA repair enzymes can correct strand breaks as they occur in the nucleus of a given cell. This is often carried out by excising the base, which is analogous to removing the center of the ladder in a DNA helix, or by nucleotide excision repair, which involves removing the center and side of the rung in the ladder of a DNA helix. In an Asgardian, these DNA repair mechanisms can be enhanced by overexpression of the enzyme poly(ADP-ribose) polymerase 3 (PARP3). This enzyme is responsible for painting parts of the genome with a sugar modification to the spool-like proteins that coil up DNA whenever and wherever damage to DNA is detected.

THE SCIENCE OF REAL LIFE

Humanity's longest-lived individual was Jeanne Calment, who lived to 122 years and 164 days. She was born to a long-lived family, lived a stress-free life, ate well, and had a regular routine for exercise. While Mrs. Calment was not quite an Asgardian, her longevity was likely attributable to the combined effects of both environmental and genetic variables. For example, in a twenty-year longitudinal study conducted by the University of Wisconsin, environmental factors such as diet were shown to affect the longevity of rhesus macaque monkeys. Individuals that had a restricted caloric intake had survived longer compared to controls that received a typical diet.

AVOID STRESS

In work carried out on healthy premenopausal women at the University of California, San Francisco, Dr. Elissa Epel found profound changes in markers for cellular senescence with psychological stress. Examining cells in blood, her group found decreased telomere length, telomerase activity, and increased oxidative damage in individuals that perceived the highest levels of stress. Surprisingly, telomere shortening in the most stressed individuals corresponded to approximately a decade of aging compared to the least stressed individuals that took part in the study.

In addition to these environmental variables, we have used the emergent profiles of chemical modifications to DNA as a measure for biological aging. Specifically, DNA methylation (see Hulk's Transformation) has been used as a molecular substrate for Dr. Steve Horvath's molecular clock. The chemical addition of methylation to cytosine residues across the genome forms unique patterns across tissues that are subject to change during the aging process. Upon examining the patterns of DNA methylation in the genomes of 8,000 human tissue samples, Horvath identified 353 sites whose methylation patterns could reveal biological age with a median error of +/− 3.6 years. The DNA methylation patterns at these 353 sites may not be the causative role behind aging, but they provide a signature that helps our understanding about how disease states can accelerate aging and mortality. While environmental factors can modulate the lifespan of a given individual, the capacity for making it past one hundred years is often a heritable difference coded within a person's genes. In humans, mutations in the genes for forkhead box O3 (FOXO3) and apolipoprotein E (APOE) have been linked to exceptional longevity in individuals who are more than ninety years old. FOXO3 is a regulator of several antioxidant proteins, suggesting it works by reducing the oxidative load imposed on tissues during aging. APOE is a protein that combines with fats to traffic cholesterol in the blood, possibly suggesting mutations of this protein may reduce cardiovascular disease risk.

Outside of what happens to a typical human population, our study of various animal models can reveal alternative strategies the animal kingdom uses to live longer. This comparative approach to understanding aging was developed by Dr. Steven Austad of the Barshop Institute for Longevity and Aging Studies. His longevity quotient is a measure of the longest-lived individual of a given species divided by the expected age of that species for its typical weight. While most animals fall neatly onto this curve, nature sometimes reveals outliers. For example, Brandt's bat can live up to forty-one years in the wild. If you compare that to other small mammals weighing 4–8 grams, this bat can live 9.8 times longer than what is expected. Recent sequencing of its genome led by Vadim Gladyshev at Harvard University revealed unique protein structures suggesting an insensitivity for several protein receptors related to growth (growth hormone and insulin-like growth factor-1 receptor).

Chapter 5

Glorious Gadgets

EXOSUITS

WHEN: *Iron Man, Iron Man 2, Iron Man 3, The Avengers, Avengers: Age of Ultron, Avengers: Infinity War, Spider-Man: Homecoming, Captain America: Civil War*

WHO: Iron Man, War Machine, Iron Monger, Whiplash

SCIENCE CONCEPTS: Human-machine interfaces, exosuits

INTRODUCTION

Humans have used millions of years of evolution to develop the biomechanics that allow us to locomote in a variety of ways. While we may intuitively decide to take a sprint or swim a certain way, this "natural" motion is not as spontaneous as it may look; in fact, all of our movements are the result of the coordinated cyclical contraction of multiple symmetric and alternating muscle groups at work. Objectively speaking, you are a machine that pulls tendons instead of actuators and pivots around joints instead of sockets. If we were to design an exosuit, we would need to consider these biomechanical move sets in order to safely orient ourselves. What technologies might we use to make an exosuit that could increase our strength and speed, while still feeling like a completely natural movement? Could we make our own Iron Man suit if we wanted? Or would we break our backs failing with faulty Hammer Tech?

BACKSTORY

There are many reasons that Iron Man's exosuit is one of the greatest inventions in the MCU. It is the product of sheer will and ingenuity, turning a middle-aged Tony Stark into an Avenger. Stark has made forty-seven mechanical Iron Man suits with a titanium gold alloy chassis and repulsors on hands and feet that allow him to fly or shoot pulses of energy. These suits come in various shapes and sizes optimized for rescues (Iron Legion) or even taking down the Hulk (Hulkbuster). One of these designs was given to Lt. Col. James

Rhodes and outfitted with a heavy ballistic arsenal to create War Machine; others were taken to make Obadiah Stane's Iron Monger, Justin Hammer's drones, and Ivan Vanko's Whiplash Mark II.

THE SCIENCE OF MARVEL

Iron Man's arc reactor functions like a betavoltaic cell that decays to release electricity in sufficient amounts to meet the energetic demands of his whole suit's infrastructure. Similarly, any fuel propulsion system used in his repulsors is analogous to the Aero-Rigs that Star-Lord and his team use. In Earth's atmosphere he probably opts for miniaturized jet propulsion using atmospheric oxygen as an oxidizer. This may have been the case for his Mark VII suit during *The Avengers*, since it flickered and lost propulsion when it traveled through Loki's portal into space. Future iterations of his suit could have learned from this lesson to implement rocket propulsion with fuel and oxidizer to sustain combustion (see Aero-Rigs).

Since fuel stores would be a limiting factor in sustaining flight, Tony Stark may have to lean into his arc reactor for ways to achieve electric propulsion. These forms of propulsion trade the high power of rocket- or jet-fueled propulsion for a more efficient use of fuel. An ion thruster, for example, would use the electricity from the arc reactor to strip electrons from a neutral fuel (noble gases such as xenon). Once they are turned into positively charged ions, the repulsion between ions is used as a propellant to generate force. Unfortunately, this form of propulsion is pretty weak and probably wouldn't be able to lift Tony Stark, let alone his Iron Man suit. Perhaps he's come up with his own cold burning superfuel that can be run off electricity.

In addition to his power of flight, we know Tony Stark has incredible strength in his Iron Man suit. What exactly allows for these feats of strength? The lifting power of his exosuit leverages the power of actuators, turning electrical energy into mechanical movement. For example, rotational actuators in his knees control the movement of his shin in a circular plane limited by a 90-degree arc. Clever engineering would have to carefully position multiple rotary or linear actuators to permit the freedom of movement enabled by a normal human musculature. In this state, Stark would need a very sensitive force-feedback system to initiate articulated movement in his suit. Without careful safeguards to match Stark's movement with his suit, pilots would injure themselves, like the unlucky fellow who twisted his spine in *Iron Man 2* using Hammer Tech.

To ensure safety, it would be ideal if the Iron Man suit could predict Stark's movements through its interfacing sensors. This predictive motion even allows Stark to pilot the suit mentally using natural human movements that fooled Spider-Man in *Spider-Man: Homecoming* and (almost) Pepper Potts in *Iron Man 3*. In the suit itself, this could be achieved by using electromyography (EMG), which measures the electrical activity in specific muscle groups; these measurements could be used to code the movements of a matching part of the suit. For example, as Iron Man carried a nuclear warhead toward the Chitauri forces in *The Avengers*, neurally encoded descending firing activated specific muscle groups throughout his body. To stabilize the load his abdominal muscles were contracting, while his arms and the rest of his upper body positioned the warhead. The coordinated EMG signals from muscle groups would correspond to the coordinated movement of linear and rotary actuators, working to turn joints or lock up his midsection to facilitate the carry. In puppet mode, this neural signal can be intercepted from the spinal cord and remotely sent to a given armor.

THE SCIENCE OF REAL LIFE

We may not have an Iron Man suit, but we have come a very long way from the marionette-like tube-connected pneumatic-actuated experimental exoskeletons of the 1960s. The earliest attempt at a practical exosuit was carried out in the 1960s by General Electric, with its Hardiman that was designed to amplify human strength by a factor of twenty-five. This joint collaboration with the US military faced many challenges related to the kinesthetic feedback that would keep a user from gripping an object too strongly. The Hardiman weighed 750 kg, which was twice the expected liftable load; there were reports that it acted violently when the legs were engaged. It was impossible to get all the moving parts to work together, and it was never piloted by an actual human when it was operational. The project was ultimately shut down in 1975 but the lessons learned contributed to GE's Man-Mate force-feedback arm.

Since then, many have engineered exoskeletons that assist with gait or help manual laborers to minimize work-related injuries. For example, Ekso Bionics has made several commercially available exoskeletons for a range of applications. Their EksoGT exoskeleton is aimed at aiding gait in individuals who have suffered a stroke or spinal cord injury. These exoskeletons are also intended to speed up recovery and rehabilitation by providing a comprehensive measurement of a patient's limited ambulation. This data allows a physiotherapist to understand exactly what parts of the suit are being driven by the patient rather than the machine. This enables a more effective physiotherapy session.

HULC

In a collaboration with Ford Motor Company, Ekso Bionics created a vest that works to alleviate the upper body strain of repetitive overhead work. Employees working on the assembly line lift their arms 4,600 times a day and up to a million times a year, increasing the risk of strains and injuries. The EksoVest provides an additional 5–15 pounds of support that can make a huge difference in a worker's day. Similar exosuits such as the Human Universal Load Carrier (HULC, get it?) have been used for military purposes and can allow a soldier to carry an extra 200 pounds, increasing their strength and endurance in the field.

It appears the future of exoskeletons will rely less on a hard metallic infrastructure and more on softer, more flexible forms. This area of research, called soft robotics, uses more compliant materials and biomimetic designs to emulate normal human muscle function. For example, the Harvard Biodesign Lab has made a soft robotic exosuit that integrates its exogenous motion in a comfortable textile suit worn by its user. Mechanical movement is overlaid onto functional muscle groups that are used in typical human walking, paralleling normal muscle movement. Batteries and motors are kept along the waist, and forces are transmitted across cables running through the textiles. The key advantage of these suits is that they allow a user to get around on various terrains, not just a hospital hallway.

NANITES

WHEN: *Avengers: Infinity War, Black Panther*

WHO: Iron Man, Spider-Man, Black Panther, Erik Killmonger

SCIENCE CONCEPTS: Nanotechnology, collective behavior

INTRODUCTION

The first signs of unicellular life appeared on Earth some 3.5 billion years ago. Only six hundred million years ago did multicellularity evolve, allowing life to develop the complex tissues and organ systems seen in today's animal and plant life. That's a lot of time and iteration to decide what to do with all these cells. Now, imagine attempting to apply that transition from unicellular life to multicellular life with machines. Given a necessary microscopic scale for these processes, it would require miniaturization into the 1–100 nanometer scale and a venture into the world of nanotechnology. Here, we will try to align our understanding of nanorobotics with the collective behavior of cells to self-assemble in order to create the dynamic "nanite" skins of Iron Man, Spider-Man, and the Black Panther.

BACKSTORY

In the MCU movies leading up to *Avengers: Infinity War*, the use of nanites allows titular heroes like Iron Man, Spider-Man, and the Black Panther to get a new look. This technology, invented by Tony Stark and Shuri independently, uses nanorobots that can self-assemble to fit a user like a second skin. Black Panther's suit has the added utility of vibranium, which can absorb and release kinetic energy. Additionally, this technology can reassemble into new shapes and forms. When facing off with Thanos, Iron Man's Mark L suit allows his forearms to turn into a shield or hammer while Spider-Man's Iron Spider can grow an extra four limbs that instinctively protect him. When not being used as a weapon, nanite-based suits act like real tissues by redistributing their subcomponents to heal torn areas of the suit.

THE SCIENCE OF MARVEL

If we were to zoom in to the base unit of one of these suits we would find ourselves looking at a nanorobotic cell. Based on the visual aesthetic given to these nanites as a liquid phase skin that wraps itself around a user, we can imagine that an individual nanorobotic cell would likely measure at most the size of a grain of sand. While there may be several features in a given nanite, they would likely possess an onboard microscopic computer capable of processing local movements and orienting with respect to other cells. While a single nanite may not be able to do much for its user, the collective behavior of millions allows it to develop emergent characteristics seen in the MCU. This is analogous to how a single fire ant cannot swim across a stream of water, but a few hundred can work together to create a raft constantly circulating individuals below and above water. At an organismal level this is mediated by animal communication and decentralized nervous systems working toward a common goal. With machines we can use wireless communication to situate other nanites in space and model and execute collective behaviors with the intuitive gestures. This would require a tremendous amount of processing power as the complexity of such computations to navigate an individual nanite would increase exponentially with the number of nanites.

In addition to sensing the organization of nanites on the surface of the skin, these nanites would have to move to disperse layers over their user's body. The ability to form two-dimensional layers could be achieved by orientable electromagnets connected to a powerful and tiny battery. One way in which this can be achieved is if each nanite has its own power source that can be used to generate powerful electromagnetic forces that create dipole-dipole interactions, which can weave into two-dimensional structures. In areas of Iron Man's armor that require the most defense, these layers can be thicker, generating the largest electromagnetic force between adjacent nanites, improving their resilience. Similarly, weaker electromagnetic forces can be exerted at the joints where less rigidity and more flexibility are required.

But what about the unique forms that nanite technology can create? One approach is to model and preform these structures with a few million nanites for a specific gesture. For example, if Iron Man is preparing to throw a particularly strong punch, he may need the added weight of a hammered fist to scrape Thanos's wrinkly chin. Since this is an intuitive reaction to a particularly strong adversary, Stark engages very little with his AI F.R.I.D.A.Y. In order to prime the Mark L suit for these types of gestures he may have used recorded data from his thousands of hours in different Iron Man suits to generate a training set for machine-learning algorithms. Over enough iterations, these algorithms would do a pretty decent job at recognizing and predicting the movements Tony makes when he needs exceptional force. This gesture, memorized in silico, would then command nanites to land a hammer-fisted blow.

THE SCIENCE OF REAL LIFE

In reality, these forms of nanorobots are particularly far from the fiction portrayed in the Marvel Cinematic Universe. However, we do have some rudimentary forms of self-assembling robots that can make two-dimensional shapes and three-dimensional shapes. These examples, however, are much larger than the science fiction nanites. Research led by Michael Rubinstein of Harvard University has shown one thousand autonomous robots can form various two-dimensional shapes. This was achieved by algorithms designed to make robots cooperate as a large group and through local interactions using 1,024 kilobots. Each kilobot, about the size of a quarter in diameter, has a vibrational motor for movement and infrared sensors to allow local communication with adjacent kilobots. The algorithm designed by Rubenstein and his team allowed for a single message to be sent to the kilobots and a resulting reorganization of their swarm into a star shape (note: this took twelve hours).

M-BLOCKS

Another example of self-assembling robots comes from MIT's Dr. Daniela Rus whose team has designed M-blocks. These small cubes can flick and attach themselves to other M-blocks, creating a modular self-assembled structure in three dimensions. Inside each M-block is a motor with a spinning mass that generates angular momentum for its movement. On the outside of each block are magnets along the faces of each cube and its edges, which allow alignment of M-blocks with one another and facilitate their movements as they flip over their edges. Unlike the kilobots, however, M-blocks take single commands wirelessly and lack the properties that implement swarm behavior or decentralized collective behavior. (It is definitely on their list of things to do!)

In addition to small self-assembling robots, the application of collective behavior on flying drones has allowed us to improve the coordination of these robots in a very organic way. Dr. Gábor Vásárhelyi at Eötvös Loránd University in Budapest, Hungary, and his team trained thirty drones to collectively swarm similar to the way birds or insects do. Unlike a preprogrammed flight that can scale into the thousands, these thirty drones were capable of monitoring various aspects of their flight and sharing it with adjacent drones to correct for any inconsistencies in the group's trajectories. This feat was achieved by running thousands of simulated flights accounting for the many variables that can affect a flock of drones. While not quite the same as nanite function in the MCU, it's a good illustration of the work that goes into nanite and nano technology: the optimization for flight of only thirty drones required fifteen thousand simulations and a supercomputer!

SPIDER-MAN'S WEB-SHOOTERS

WHEN: *Spider-Man: Homecoming, Captain America: Civil War, Avengers: Infinity War*

WHO: Spider-Man

SCIENCE CONCEPTS: Biotechnology, genetics, synthetic biology

INTRODUCTION

Spiders make up a diverse and successful group of arthropods with over 45,000 different species that we know of. Each species of spider has developed many traits that have led to their success across habitats (deserts, jungles, under your bed, etc.). However, spiders are probably best known for their unique ability to spin silk from one to four pairs of glands called spinnerets located on the bottom of their abdomen. These spinnerets can be used to make trail lines and sticky silk for webs or a finer silk to wrap up prey. These silks are made of protein and are often considered stronger than steel by weight and more flexible than rubber. So, how exactly could we take advantage of these silks to do whatever we want, especially if it involves swinging across Manhattan's skyline?

BACKSTORY

After getting his fateful spider bite, Peter Parker engineers his own spider silk with web fluid contained in capsules fastened to his wrists. He claims his web fluid was made from household supplies and whatever he was able to scavenge from his high school chemistry lab. These web-shooters allow Spider-Man to swing through New York City by shooting draglines between skyscrapers, letting gravity and momentum do the rest of the work. Much like a spider, Spider-Man also uses his web-shooters to snare, bind, and blind criminals for the local police to arrest. Once spun, this thread can dissolve within two hours, leaving no trace behind.

THE SCIENCE OF MARVEL

In Spider-Man's first introduction to the MCU, Tony Stark praises Peter Parker's web fluid for its extreme tensile strength. This tensile strength, or the resilience of a material during a pulling stress, paired with its light weight and flexibility make it an incredibly useful tool in the web slinger's arsenal. All things considered, it is pretty difficult to imagine how Peter Parker could have made his web fluid from the resources available to a high school student (if this was the case we would have more teenagers swinging through Manhattan). Perhaps the easiest explanation for Spider-Man's web-shooters is to assume Stark Enterprises fronted some of the resources to build his upgraded Spider-Armor MK II suit and the Iron Spider.

Under Stark's guiding hand, the most logical place to synthesize spider silk would be from spiders themselves. Ancestral spiders began evolving their silk about 419 million years ago, and they have had plenty of time to perfect it for different uses. Luckily, a summary of this lengthy process can be found within the information encoded in spider genomes. Here, we can find chemical blueprints for the proteins that make different spider silks. Given the various species of spiders that have tailored their silk (if you pardon the pun), the MK II's augmented reality interface should *at the very least* provide 576 possible combinations of webbing.

When one considers the spider silks as proteins, we categorize them in a class of "structural" proteins. A structural protein like spider silk is a building block that imbues mechanical properties at the molecular level to a material super fiber. A good example of this is your hair, which is made up of a structural protein called alpha-keratin that is produced in hair follicle cells. At a molecular level, alpha-keratin is a protein that folds into cylindrical helices. These proteins further assemble to form fibers aggregating into supercoiled structures that eventually become a thread of hair. In the case of spider silk, "spidroins" can be substituted for alpha-keratin, with hair follicle cells being replaced by gland cells in spinnerets. Spidroins are very large proteins and when aggregated they generate intermolecular forces that increase their tensile strength. Consider that alpha-keratin is made of approximately 500 amino acids whereas spidroins average around 3,600 amino acids.

Which spiders would make the best webs for Spider-Man? Stark Labs probably draws inspiration from the iconic spiral wheel-shaped webs that are prominently featured on Spider-Man's chest. These are created by orb-weaver spiders. From these spiders you can use dragline silks to provide the tensile strength Spider-Man needs to swing his body weight between buildings (or, as seen in *Spider-Man: Homecoming*, to hold together a ferry that is being split in half). In order to trap criminals in flight, we might use an orb-weaver spider's sticky flagelliform silks, which make up the capture spiral of orb webs, or the stiff aciniform silks, which are used to bind and store prey. While most spiders produce variations of these silks, it is safe to assume that we can exploit this property to procure super specialized versions of these silks for different uses. For example, for particularly strong draglines, we could turn to Darwin's bark spider, which is known to make the largest orb webs in the world (up to 30 square feet).

THE SCIENCE OF REAL LIFE

While we may not be making our own web-shooters anytime soon, we have been able to parse out the genes responsible for making different spider silks, introduce them transgenically into other animals, and even promise some commercial goods for the coming decades. Unfortunately, this type of technology takes a lot of time and requires interdisciplinary approaches to picking the right spidroins, finding the appropriate "biological factory" in which to produce these silks, and deciding on which applications for such a technology are the most practical and feasible.

Since the spidroin genes were isolated and sequenced in the 1990s and early 2000s, we have made strides in understanding both the genetic and biophysical properties of spider silks. For example, we had determined only part of the sequence of a single dragline silk in 1990 but by 2017 we had sequenced the entire genome of the orb-weaver spider, *Nephila clavipes*, while cataloging twenty-eight different spidroin genes and 394 variations of the motifs in spider silk proteins that define part of their mechanical properties. This wealth of information and the decreasing cost of sequencing technologies has allowed us to turn the average cost of a billion dollars and twenty years of work into an overnight experiment that costs under a thousand dollars. This is particularly important since it allows researchers to examine the differences between spider species with more interesting silk-spinning strategies.

Naturally, the best animal to collect spider silk from would be the spider itself (after all, we harvest silk from silkworm cocoons for textiles). However, it is difficult to farm a good yield from such small animals and most spiders aren't social enough to be housed in close quarters of each other (they will eat each other). To circumvent these issues, many have tried to transfer spidroin genes into farmable animals or crops. This approach finds synergy with our ways of raising domesticated animals and crops to produce large yields of protein. Spider silk genes have already been transferred to bacteria, tobacco, potatoes, silkworms, and goats. While the first "spider goat" was made in Canada by Nexia Biotechnologies Inc., research currently being carried out at Utah State University has filled barns with goats carrying spider silk proteins. Similarly, other companies, including Kraig Biocraft Laboratories, have sought to exploit the established sericultural production of silk in silkworms to produce spider silk for commercial needs. Once produced in reasonable amounts, spider silks can then be harvested, purified, dried, and spun into fibers to be used in textiles.

So, aside from its utility for Spider-Man, how can we use spider silk? Currently, spider silks have shown potential for various applications in biomedical, textile, and cosmetic industries. For example, AMSilk supplies spider silk protein in powder, microbead, and hydrogel form to retain moisture and breathability in cosmetics products. Similarly, they have also proposed the use of spider silks as hypoallergenic material for bandaging burns or coating implantable medical devices. AMSilk and Spiber Inc. have also marketed the use of spider silks for textiles by partnering with North Face and Adidas to make shoes and parkas. So while you may not be using spider silks to swing on skyscrapers anytime soon, there's a good chance you may be able to wear them in the coming decades!

FALCON'S REDWING

WHEN: Ant-Man, Captain America: Civil War, Avengers: Infinity War
WHO: Falcon
SCIENCE CONCEPTS: Optics, radio waves, artificial intelligence, holography

INTRODUCTION

The ability to scout a terrain for dangers is necessary for a military force to be efficient on the field. Good reconnaissance tells of dangers in the field and allows for the best tactical decisions to neutralize a target. The earliest forms of reconnaissance used mounted cavalry (due to their mobility in the field) and today various technologies are used. For example, high-resolution multisensory cameras, satellites, and drones have all added new layers to the intelligence that can be collected by a military force. Specifically, unmanned aerial vehicles (drones) are widely used for reconnaissance to track the movement of enemies and identify civilian populations within the field. What tools can we implement drones with to provide them the best edge in the field, and would they ever be as effective as Falcon's Redwing?

BACKSTORY

Sam Wilson, or Falcon, has his own personal winged exosuit, hand-mounted machine guns, and an aptitude for acrobatic maneuvers. Despite being an expert combatant, Wilson's greatest utility is the use of his drone, Redwing, and a pair of flight goggles to comprehensively surveille the field. In *Captain America: Civil War*, Wilson uses these tools to see through the walls of the CDC building in Lagos to identify the positions of all armed hostiles. Using his goggles he is also granted vision beyond the visible spectrum and the ability to see things that are both very far away and incredibly small. This was seen when his microscopic view of Ant-Man allowed him to notice and track the moving miniaturized target throughout the Avengers headquarters.

THE SCIENCE OF MARVEL

In the events of *Captain America: Civil War*, Redwing was able to navigate quietly to a fixed position, where it identified the location of Brock Rumlow (Crossbones) through the walls of a building. If Redwing could use wireless frequencies and high-resolution sensors, it could likely use this technology to detect motion, shape, heart rate, and breathing patterns of a target and match these stats to the biometric signature of the target, in this case Rumlow. The first step would require using wavelength frequencies longer than the obstacle they are being transmitted through (Wi-Fi waves, for example, are ~76–127 mm, and radio waves are 1 mm–100 km). These waves could be transmitted and received by multiple antennae on Redwing to create a composite three-dimensional image of a scene. Similar to the way a hologram is created, a set of scanning antenna can move through hovering motions while a reference antenna scans reflected radio frequencies in a stationary position (perhaps deployed at a fixed position). The returning wave form and the interference between reference and scanning waves gives information on the texture and distance of different objects.

This, however, only gives us a fuzzy blob of an object behind a wall. One way to clean up the signal is to deconvolute the hologram using the point spread function (PSF) of light to the reconstructed radio-frequency hologram. We might conceptualize how the PSF works for this type of application by imagining the data collected by Redwing as a two-dimensional "slice" of a scene behind a wall. If the image was 100 percent accurate it would look identical to a cross-section of the physical scene; however, this is not what we typically see. Data collected through a radio frequency can come back as poorly defined as clouds of movement due to artifacts introduced by the radio frequency waves. These artifacts, as with visible light, can be seen as issues of contrast, saturation, or definition. If you take into account these artifacts and the math and optics that can generate them, you can computationally subtract this noise from your final cross-section. By calculating the PSF of a two-dimensional image stack that makes up the radio-frequency hologram, we can resolve a clearer picture of the hidden environment.

IDENTIFYING AN INDIVIDUAL

Would it be possible, using this technology, to identify Rumlow based on his gait and voice? A voice identification would be easy enough since S.H.I.E.L.D. databases would already have vocal information on file for Rumlow. This could be achieved by digitizing the sound waves captured on radio communications and parsing frequencies of Rumlow's voice into different bands. These isolated parts of his speech are called formants; they form the basis of a vocal fingerprint. Any video footage of Rumlow's gait and typical movement could be used to train a convolutional neural network to create classifiers that recognize his gait. A convolutional neural network uses spatial information within an image to be abstracted through a neural network into a new layer of data that is subject to the same iterative abstraction. This form of spatial abstraction is called a convolution.

THE SCIENCE OF REAL LIFE

In real life, each of these aspects of Redwing's technical capabilities are completely within the realm of possibility. However, to integrate so many tools into one hovering drone is a challenge. The biggest limitation is on the processing power needed to remotely create a holographic volume, deconvolute it, feed this data into a convolutional database to identify an agent, and to have this take place through three stories of a building. The idea of using long wave frequencies on the electromagnetic spectrum to detect motion was actually implemented in handheld radars used by first responders.

One example, the Range-R, is analogous to a motion detector that can be used through walls to identify trapped civilians or to manage hostage situations. More recently, researchers led by Dr. Dina Katabi at MIT have built upon this concept to develop several new tools that use Wi-Fi and radio frequencies to measure motion and defined shapes in a walled room. Using Wi-Fi, her group was able to tune two antennae to produce waves that would cancel themselves out in a static object (a wall, a door, or a chair) and reflect waveforms that could be detected in a moving object. This creates a wave tracing a subject's movements forward and backward over time.

Her group also used neural networks to pair annotated video footage of a walking person with their resulting radio-frequency reflections. This allowed an artificial intelligence to abstract the modular motion of a human into a stick figure that could represent human motion in front of the radio frequencies. When that same person walked behind a wall, the neural network was able to continue emulating the motion of the virtual skeleton. This research, in addition to providing an obvious benefit to first responders, may also allow better surveillance and health monitoring of different patient risk groups, such as people who suffer from Parkinson's disease. These neural networks were even able to reidentify an individual blindly, only from the person's gait, with about 83 percent accuracy. A separate group from Germany, led by Dr. Friedemann Reinhard, used Wi-Fi to create a system similar to the holograms we speculate for Redwing. Reinhard's group used a fixed antenna and a scanning antenna to generate coherent two-dimensional waveforms that can provide three-dimensional data of all the objects one wave interacted with. This approach, analogous to holography, was able to scan a cubic meter and simulate this process for a larger (10 cubic meter) building.

Approaches to deconvolution are also commonly used for interpreting images taken of planets incredibly far away, as well as very

CHAPTER 5: GLORIOUS GADGETS

small microorganisms. Using typical properties of visible light we can apply a variety of algorithms to mathematically improve the resolution of an image. Many of these processes have been recently implemented with neural learning networks that use training sets of other images and the optical artifacts they generate to reduce the noise they contribute to an image. As described earlier, such approaches can also be used on multiangle video acquisition of a person's gait. A group led by Noriko Takemura from Osaka University in Japan used this information and convolutional neural networks to measure gait for forensic identification.

AERO-RIGS

WHEN: *Guardians of the Galaxy Vol. 2, Avengers: Infinity War*	
WHO: Drax, Rocket, Gamora, Star-Lord	
SCIENCE CONCEPTS: Turbines, rocket propulsion, jet propulsion	

INTRODUCTION

Gunpowder fueled the use of explosives and projectiles for warfare in the tenth-century Song Dynasty and during the Mongolian invasions of the thirteenth century. Advances in the military use of rockets continued, and, in 1918, we began to use liquid propellants to fuel rockets. This innovation led to the possibility of manned spaceflight and the first fictitious depictions of jetpacks in 1920s pulp comic books. These inventions and their fictitious renderings inspired the imaginations of DIY engineers that would make the world's first rocket belts and turbine-powered jetpacks. What does it take to defy gravity the way only a rocket can? Can the Aero-Rigs used in *Guardians of the Galaxy Vol. 2* exist?

BACKSTORY

At the beginning of *Guardians of the Galaxy Vol. 2*, the Guardians find themselves face-to-maw with the energy-consuming Abilisk. To maneuver around its tentacles and rows of teeth each member of the team equips himself or herself with Rocket's newly built Aero-Rigs (except Drax because he has sensitive nipples). These one-size-fits-all rocket packs expand from a small disk that fits like a vest over their users, allowing them to generate controlled rocket propulsion. When in use, the Guardians can navigate these Aero-Rigs with six degrees of freedom by making small corrections to their flight path with their body's intuitive movement. This Aero-Rig is used in the final escape from Ego where Yondu sacrifices himself to save Peter Quill, carrying them both out of Ego's atmosphere.

THE SCIENCE OF MARVEL

In *Guardians of the Galaxy Vol. 2*, each Aero-Rig is made from a snug-fitting vest attached to two rocket thrusters, which are constantly pivoting to thrust in the direction its user wishes to travel. Pitch, yaw, and roll are immediately sensed by an onboard computer and sensors on the vest, causing the two rocket thrusters to adjust to the user's movement. Aero-Rigs are also capable of escaping the atmospheric pressure of Ego into the vacuum of space, suggesting they need to carry their own combustible fuel.

Rocket propulsion requires an oxidizer and a fuel to create combustion. Both fuel and oxidizers are kept supercooled in an Aero-Rig to make them as dense as possible, allowing the fuel to be used more efficiently. Combustion in the rocket occurs by combining the oxidizer with fuel in a combustion chamber and directing its path through a converging, then diverging, space. Starting in the combustion chamber the combination of oxidizer and fuel creates a chemical reaction, causing a resulting high pressure. This pressure then gets passed through a smaller channel, called the throat, and into the nozzle, which fans out. Inside the nozzle pressure is lower but the speed of the gases exiting the combustion chamber and throat is much higher. This entire process creates thrust in the opposite direction of the exhaust that exits the Aero-Rig. This is Newton's third law in action, demonstrating that with each action there is an equal reaction.

How exactly would Star-Lord's or Gamora's movements send information to the rocket engines to adjust flight paths accordingly? There is likely a powerful onboard computer with several sensors that record a user's movements and translates them to changes in the angles of both thrusters. Sensors in the Aero-Rig must include a combination of gyroscopic sensors that can measure the angular velocity of Star-Lord as well as tilt sensors that are similar to a carpenter's level. Microelectronic mechanical systems (MEMS) would be favored due to their size. They can be placed in different orientations in the Aero-Rig to detect different angular velocities on different axes. MEMS work by housing a small object such as a piezoelectric crystal or ceramic on a spring where an electric current can be run through it to cause vibration. Depending on how this vibrating object interacts with fixed plates in its housing, it will communicate a characteristic of the angular velocity to the onboard computer. Tilt sensors, on the other hand, work by having multiple electrodes enclosed with an electrolytic fluid and a bubble in a small dome. The electrolytic fluid has resistance that changes with the level of the fluid over these electrodes, providing an angle of tilt relative to gravity for an Aero-Rig. These sensors, placed in multiple places on the Aero-Rig, would provide a comprehensive view of Star-Lord's movement that can be processed in silico and used to change the direction of rocket thrusters through continuous feedback. Since MEMS do not rely on gravity and tilt sensors do, an Aero-Rig would continue functioning in space due to MEMS and combustible rocket fuels.

THE SCIENCE OF REAL LIFE

In real life, we have had jetpacks for quite some time. While these jetpacks have not been made by a super intelligent raccoon, they have been getting better with each year. The earliest version of jetpacks was invented in 1919 by Aleksandr Fyodorovich Andreyev, a Russian inventor. This prototype was fueled by methane and oxygen rockets and had a 2-meter wingspan. In 1958, Thiokol Chemical company marketed its jump belt as a means to allow wearers to use nitrogen tanks to increase their ability to jump. In 1960, Bell Labs engineered a jetpack that employed the use of pure hydrogen peroxide (H_2O_2) with a silver catalyst to create a superheated exhaust of steam and oxygen ($2H_2O_2- > O_2 + H_2O$). The main disadvantage of this approach was that it used up fuel incredibly fast, limiting flight times to about thirty seconds.

Given the many disadvantages of using H_2O_2 for propulsion, teams turned to turbojet engines that used traditional kerosene for fuel. This led to the invention of the WR19 by Bell Aerosystems through funding from the US Defense Advanced Research Projects Agency (DARPA). This prototype used gas turbines and was able to generate up to 1,900 newtons of force while only weighing 68 pounds. In test flights carried out in 1969 the WR19 could reach 7 meters in altitude and speeds of 45 km/hr with flights lasting up to five minutes. Unlike previous models, the WR19's gas turbines worked like the jet engines of planes and used propeller turbines to concentrate airflow into two streams. One stream gets compressed by smaller fans into a combustion chamber where it mixes with fuel to create combustion. The resulting exhaust exits the turbine by passing through another propeller connected to the inflowing propeller, allowing residual energy to speed up the flow intake. The second stream of air flows around the engine to cool the exhaust temperatures.

The closest thing to the Guardians' Aero-Rig may be NASA's simplified aid for extra-vehicular activity (EVA) rescue (SAFER) system. SAFER is an emergency jet propulsion system for astronauts to use during spacewalks if they become untethered from their vehicle. This simplified manned maneuvering unit uses gas ejection to push an astronaut in a controlled way at zero gravity.

JETPACKS TODAY

The most impressive existing jetpack belongs to JetPack Aviation and is called the JB11. The JB11 has a six-turbine engine and can reach speeds of up to 240 km/hr with automatic flight stabilization safety technology. These jetpacks allow you to change altitude and yaw but are limited in terms of roll or pitching too far ahead just as Rocket does in his Aero-Rig. Additionally, many of these jetpacks require the use of both hands to regulate altitude and yaw.

Many of these jetpacks use a variety of the sensors described earlier but probably lack the degree of sensitivity that Rocket's jetpack has. At best, fine control using such sensors is limited to one plane of movement, similar to the way in which a Segway scooter can detect the balance of its user but can only navigate forward, backward, left, and right. Measuring the electric potential of muscle groups in the chest may also provide some read on tentative human movements, but it would be a leap to integrate all this information with sensors and turbine engines to come up with something like the Aero-Rig that Rocket designed.

Chapter 6

Aggressive Assaults

HULK'S THUNDER CLAP

WHEN: *The Incredible Hulk*

WHO: Hulk

SCIENCE CONCEPTS: Physics, biomechanics, acoustics

INTRODUCTION

Our eardrums perceive sound by detecting the vibrations of pressurized air coming from a given point in space. Since these vibrations of pressurized air act like waveforms, we can use concepts such as frequency and magnitude to improve our understanding of how we perceive sound. For example, the amplitude of a sound wave corresponds to the loudness of that sound and the frequency of a sound wave can correspond to the pitch of that sound. These physics of sounds determine the ranges and limits of human hearing—how we can sense something as quiet as a pin drop or as loud as a cannon fire. Is it possible to weaponize sound or at least calculate the force of a clash based on the sounds produced?

BACKSTORY

In the events of *The Incredible Hulk*, the final confrontation between the Abomination and the Hulk takes place in the streets and on the rooftops of Harlem. In their first clash, they both jump toward each other with so much force that their impact resonates down 125th Street, causing all the nearby car alarms to go off. Up above in a military helicopter, General Thaddeus Ross and Dr. Elizabeth Ross attempt to aid Hulk by raining gunfire onto Abomination. In retaliation, the Abomination pulls down their helicopter, causing it to leak fuel and catch fire on a nearby rooftop. Knowing he must save them, Hulk claps his hands together with so much force that he creates a rush of air that extinguishes most of the fire engulfing their helicopter. What just happened and how?

THE SCIENCE OF MARVEL

The Hulk brings his hands together with flat palms facing each other to create the largest sound he can make. As these hands come together, they compress air between them, causing a *crack* that's immediately dampened by the elasticity of the flesh in his hands hitting each other. If a typical human were to do this, the sound of this clap would be heard by someone 343 meters away, one second later (the speed of sound). However, when the Hulk does his thunder clap, the muscles in his chest and arms are pulling each arm together, accelerating to at least 171.5 meters per second at impact, or half the speed of sound. At this instant the Hulk has compressed air quicker than it can radiate as a sound wave. The compressed air, or wavefronts, lag behind the source and create a sound barrier with very high amplitude. This leads to a shock wave that will hit the helicopter approximately 10 meters away with a sonic boom. When the shock wave's speed decreases over distance, it returns back to a normal sound wave of compressed air. As the sonic boom passes the helicopter, it trails negative pressure after the compressed wavefront, which is filled with the ambient air behind its source, causing a gust of wind that wicks off the remaining fire and fuel engulfing the helicopter, putting it out.

THE SCIENCE OF REAL LIFE

How dangerous is sound in real life? Well unlike the Rosses inside that helicopter, anyone caught in a sonic boom would probably suffer burst eardrums and a total loss of hearing. In fact, the weaponization of sound has been used to deter large crowds and provide non-lethal forms of crowd control. The magnetic acoustic device made by California-based HPV Technologies can generate a targeted beam of sound over a distance of more than a mile using planar speakers; it can cause a great deal of discomfort in people in its path. Similarly, the Israeli military has developed a targeted sonic blaster, called the Scream, that can disrupt the function of the inner ear and make targets nauseous and dizzy. Such devices have even been used to deter teenagers from loitering and committing vandalism. For example, the Mosquito was designed to create 17-kHz frequency sounds that can only be heard by individuals between thirteen and twenty-five years of age. Most of these technologies play with the design of a speaker and increase the frequency and amplitude of sound waves to give them more energy and intensity, respectively. A typical speaker is made with a circular magnet. An outer magnet has one polarity, and the inner core has the reverse polarity, and the inner magnet has a coil of copper wire nested between both polarities. When an alternating electric current is run through this copper wire, its interaction with the magnet and its inherent electromagnetisms will bounce the coil up and down, moving a flexible plastic diaphragm. This diaphragm and its movements will compress air at specific frequencies and amplitudes, generating the sound you hear in your earphones, speakers, or phone.

Is there any life-form on this planet capable of its own thunder clap that can break the sound barrier? In tropical and temperate waters around the world exist small crustaceans that, for their size, can put the Incredible Hulk to shame. These pistol shrimp can grow up to 5 cm and belong to the family Alpheidae. They have one regular-sized claw and a second larger one. This larger claw has a cocked slip joint that can build a great deal of tension.

This claw snaps so quickly that it creates a jet stream of low pressure that pulls gases out of the surrounding water to create superheated bubbles that expand. This cavitation bubble then immediately succumbs to the pressure of the surrounding water, causing it to collapse at very high pressure and temperature at around 7,232°F (4,000°C). Reminder: water boils at 212°F (100°C). This death bubble is used to stun prey, knocking them out for the pistol shrimp to eat. To be fair, it should be noted that sound travels much faster in water than it does in air. So where the speed of sound in air is 343 meters/second, in water it is 1,498 meters/second at the same temperature. In tropical waters a higher temperature means sound travels even faster, giving the pistol shrimp more bang for its effort and a lower threshold energy to break the sound barrier.

THOR'S LIGHTNING

WHEN: Thor, Thor: The Dark World, Thor: Ragnarok, The Avengers, Avengers: Age of Ultron, Avengers: Infinity War
WHO: Thor
SCIENCE CONCEPTS: Electrostatic discharge, meteorology, electricity, lasers

INTRODUCTION

If you were to rub your feet on a carpet and poke someone, you would realize that you are capable of discharging small amounts of electricity from static buildup. The fundamental principle involves electrons traveling away from a negatively charged object to a positively charged object. Normally this does not occur too often, and when it does, it occurs in the form of a spark between an outlet and a plug. However, an electrical discharge can increase its effects as it scales in size in clouds and in thunder storms. What exactly causes lightning and thunder? What happens if you are struck by lightning? And is it possible to harness these elemental forces and use them the way Thor can in the Marvel Cinematic Universe?

BACKSTORY

Thor is capable of harnessing one of the most powerful elemental forces observed on Earth, lightning. In his earliest adventures leading up to *Thor: Ragnarok*, he relied on his mystical hammer, Mjolnir, to amplify his connection to elemental forces that allow him to generate lightning and thunder. After the worst sister in the Nine Realms breaks Mjolnir, he is forced to reconnect with his elemental control over lightning. Without any mystical weapons and after a beatdown from the Hulk on the gladiator planet of Sakaar, Thor finds this power within himself and reclaims his title as the Asgardian God of Thunder. When controlling his electrical powers, Thor uses his own body as a conduit for creating lightning strikes while also emanating several forms of electrical discharge.

THE SCIENCE OF MARVEL

When Thor brings down lightning he is effectively releasing energy from the heavens to Earth. As clouds form above Thor, an updraft of supercooled droplets and small ice crystals rise upward, colliding with soft hail in the center of a cloud mass. These collisions result in the exchange of electrons, generating a positively charged area in the upper strata while creating a negatively charged area in the middle of the cloud mass. Since the collection of dust, air, and ice that make up a cloud act as excellent insulators, a tremendous amount of energy is required before any electricity can be discharged. When electrification reaches a peak at which a cloud simply cannot insulate (this is referred to as the dielectric strength), these electrons will be discharged in the form of lightning. Given Thor's astral origins from Asgard, we can speculate that he may use some control over cosmic rays arriving from distant exploding stars, capable of creating a conductive path for a lightning strike.

The electrical discharge that forms lightning more often occurs within or between clouds. However, since Thor is standing 12,000–21,000 meters below, he uses Mjolnir as a lightning rod to ground an incoming lightning strike (or leader) to the positively charged ground. At about 46 meters from the positively charged Mjolnir, electricity travels upward through Thor and his hammer, generating a streamer that touches the downward leader. In this instant all the excess negative charge of the cloud and positive charge of the ground equilibrate through the transfer of up to 100 billion volts across a lightning network. This lightning heats up the surrounding air to 54,000–90,000°F (30,000–50,000°C), causing air to expand and ionize. The ionization of air will generate a distinctive "clean" smell through the generation of ozone and the expansion will generate thunder as a shock wave within a 10-meter radius around Thor before it becomes a sound wave cracking at 343 meters per second in the surrounding vicinity. Depending on the nature of cloud formations above him, Thor is probably able to summon lightning from up to 160 kilometers away.

THOR WITHOUT HIS HAMMER

What happens if you take away Thor's hammer, Mjolnir? Mjolnir was only ever meant to help Thor focus his powers (quite literally, actually). In *Thor: Ragnarok*, Thor learns to generate electricity from his own body, causing it to discharge in the surrounding air. This means he produces so much electricity that it is capable of escaping through the insulation of air, which has a dielectric strength of 3 kV/mm! This allows him to dissipate large amounts of electricity from his body in a somewhat focused chain that can conduct through multiple positively charged enemies. Thor is able to take advantage of this fact in *Thor: Ragnarok* when all the metallic armor and swords of Hela's Berserkers are used to conduct chain lightning from Thor's bolts in his final confrontation on the Bifrost Bridge.

THE SCIENCE OF REAL LIFE

What exactly happens if you stand with a hammer under an incoming thunderstorm waiting for lightning to strike? If you are lucky, nothing. Unfortunately, a human does not stand up to the power of a lightning strike the way an Asgardian god does. In the United States, about 10 percent of people struck by lightning die, making up fifty deaths every year. Within the first few milliseconds of being struck, you will likely form third-degree burns at the entry and exit areas of the lightning strike. The heat and flash of lightning can cause blood vessels to rupture, possibly forming

Lichtenberg figures along your tissues. Lichtenberg figures trace the path of the electrical discharge as it passes through insulating tissues and look like fernlike patterns on your skin. In addition to superficial damage to tissues, the lightning strike may also disrupt the rhythms of your heart, lungs, and brain. Most deaths caused by lightning are due to cardiac arrest, respiratory failure, and various neurological disorders ranging from mood changes to seizures. The shock wave from lightning will likely burst your eardrums as well as tear your clothes up with a force capable of throwing you away from the shock wave, causing you to suffer from additional fall-related injuries (broken bones, internal hemorrhaging, etc.). Needless to say, you would not look as composed as Chris Hemsworth does when he swings his hammer around.

What about our own ability to generate lightning using human technology? Surprisingly, we have developed tools to control aspects of weather in order to subdue dangerous weather conditions or even as a weapon to attack enemy nations. In order to minimize the hazards posed by cyclones, the National Oceanic and Atmospheric Administration has suggested that lasers can be used to discharge lightning from storms that are likely to become hurricanes. In practice this technology relies on using a laser beam to create a "tunnel" of ionized air. This high-energy pulse can shake off electrons in its path, creating a stream of plasma and resulting in a linear path of positively charged air. This path acts like a tunnel to offer a path of least resistance for lightning or a high-voltage source to guide an electrical discharge. Researchers from Teramobile, a collaboration between research institutes in France (CNRS) and Germany (DFG), have put this idea to the test by shooting short laser pulses into the thunder clouds above Langmuir Laboratory for Atmospheric Research in New Mexico. Using their mobile laser, this group was able to initiate and record electrical discharges within the synchronized laser pulses. Other approaches include using a rocket attached to a copper wire filament that can be shot into a thunder cloud. Researchers at the University of Florida have used this technique, causing a lightning leader to conduct electricity through the wire filament, exploding the wire and creating a column of ionized air that tunnels remaining lightning strikes. In a similar fashion, the American military has made laser-guided lightning weapons, called laser-induced plasma channel (LIPC), that create a virtual tunnel through which a high-voltage source can predictably travel. As weapons, LIPC can be used to take down enemy vehicles as well as detonate unexploded ordnance that may be a hazard for infantry and vehicles.

BLACK WIDOW'S BITE

WHEN: *Iron Man 2, The Avengers, Avengers: Age of Ultron, Avengers: Infinity War, Captain America: The Winter Soldier, Captain America: Civil War*

WHO: Black Widow

SCIENCE CONCEPTS: Electric capacitance, Tasers

INTRODUCTION

We typically use electricity to power technology to various specifications. In the United States, most appliances run off of a 120-volt outlet but others require a larger potential of up to 240 volts, such as electric cooktops. Similarly, the typical amperes (amps) of current in a plug can go up to 15 and will draw electricity differently depending on the wattage required by a given appliance. If you don't have a continuous source of electricity, what is the highest voltage that can be stored in a battery and how could it be weaponized to create effective Taser guns? If you were to get shocked by one of these Tasers, what would happen to your body and what would you feel?

BACKSTORY

Black Widow is an expert combatant who knows how to acquire all of her weapons in the field. However, if given the time she prefers using her electrified gauntlets, batons, and throwable disks to even the odds against multiple enemies. These weapons give her the ability to stun an enemy with a tase, or what is better known as the Widow's Bite. In *Iron Man 2*, she uses electrified disks to take out several of Justin Hammer's guards on her way to Ivan Vanko's control room. During *Captain America: The Winter Soldier* and the hostage rescue on the Lemurian Star, she uses her gauntlets while grappling armed guards and shocking them into submission. Lastly, her dual batons and electrically charged suit have repeatedly come in handy; she uses them against Ultron's Iron Legion, Thanos's Black Order, and an army of Outriders.

THE SCIENCE OF MARVEL

Natasha Romanoff's electrically charged weapons deliver an incapacitating blow to any enemy unlucky enough to stand in her way. But what exactly does an electric shock do to your body? At one extreme you could get struck by lightning (see Thor's Lightning), but we know that the Widow's Bite is a measured pulse meant for mostly non-lethal intervention (when she isn't taking out Ultrons and Outriders). Assuming that the stingers she uses on her gauntlets were stabbing you in the neck, you'd probably receive a pulse that would seize all of your muscles. This is because your muscles are innervated by the nervous system that controls your motor movements through an electrical current. Normally, if you wanted to do an arm curl, your brain would send a descending electrical current to the musculocutaneous nerve in your arm, causing the biceps to contract. However, if you were shocked by Black Widow's gauntlets, the current running through your body would cause many of these neurons to fire and contract your muscles at the same time, causing your whole body to seize.

Black Widow's weapons can shock as long as the target is completing a circuit for electrons to flow. As she punches with her gauntlets, two electrodes make contact on an enemy, completing a current that travels at high voltage through her opponent. The effects of her electric shock vary depending on three things: their current, voltage, and the resistance of the target. Current is represented by the flow of electrons traveling between electrodes, and the voltage is the potential difference of the electrons traveling between electrodes. The resistance of a circuit determines how much of the current can pass through the electrodes and is represented by Ohm's law, in which resistance = voltage/current. In Black Widow's case this depends quite a bit on her target. For example, in the early events of *Captain America: Civil War* she repeatedly tried to shock Crossbones but failed due to the resistance of the scar tissue all over his body and the armor he was wearing. One way she could circumvent this would be to increase the voltage of her shock to decrease the resistance and allow for a more effective shock. Romanoff probably also needs to avoid shocking herself since she is always in close quarters with whoever she is fighting with. Her own black suit is probably made from an insulator such as rubber.

The ways in which she can use her sting limited her use of available power supplies. For example, during *Avengers: Age of Ultron*, she trades in the all-black bodysuit for one that has a blue stripe that can store electricity outside of her gauntlets. This is partly shown when her own kicks shock Strucker's army in the Sokovian stronghold. To sustain the repeated use of her electricity-based weapons, Black Widow can insulate her suit with layers of smart textiles; these can behave as a capacitor that can store electricity. A capacitor generates an electric field between an anode and cathode without contact and can temporarily store an electric charge. This can be integrated into textiles by inking a layer of fabric with nanomaterials that can have it behave as an anode or cathode that is separated with an insular fabric.

THE SCIENCE OF REAL LIFE

Since Romanoff's Widow Bite is based on several readily available technologies, it isn't too far a stretch to imagine that her sting is based on how a Taser works or to compare it to new research into smart textiles. Tasers are widely used by civilians and law enforcement agencies around the world. Like Black Widow's gauntlets they deliver an incapacitating shock by conducting up to 50,000 volts at a very low current, but they do so with gas-propelled barbs. This high voltage is used to overcome the resistance of clothes that would impede the circuit from completing and the resistance that would come with it. Similarly, this low current is used in order to exert non-lethal force and subjugate a criminal.

In a typical Taser design, a 9-volt battery is used to generate a 50,000-volt shock using a capacitor. This capacitor builds an electric charge through the sustained current of the battery running electrons across two electrically conductive plates separated by a dielectric material insulator. Over time the capacitor collects a charge by gaining electrons at one plate, generating an electric field between the two plates and losing electrons at the other plate connected to the remaining circuit. A high-voltage capacitor can store a high charge by increasing the surface area of the plates, decreasing the distance between the plates, or by increasing the permittivity of the insulating dielectric material (i.e., the ability to store an electric field).

STUN BATONS

Black Widow's stun batons are also probably analogous to stun batons used for managing livestock in the agricultural industry. However, the shock delivered by a stun baton is different from Tasers. Compared to a Taser, the voltage differential is less but the current is higher in order to deliver a painful shock that can "motivate" an animal to move between two points. With the Widow's Bite, it is particularly important to modulate the amount of voltage and current traveling through a circuit for different purposes. For example, a high-voltage/low-current shock can knock out a target, whereas a low-voltage/high-current shock can be used for interrogations.

CHAPTER 6: AGGRESSIVE ASSAULTS

The research on smarter textiles is also an exciting prospect for the future of functional clothing. In a collaboration between the Shima Sheiki Haute Tech Lab and Dr. Yury Gogotsi at Drexel University, various iterations of Black Widow's blue-striped suit are being made in their prototypical forms. Specifically, this group is creating yarn electrodes for wearable supercapacitors that can be knitted into garments. The electrodes of this yarn are made from carbide-derived carbon (CDC), which are carbon atoms bound to less electronegative elements; they can be synthesized to create porous and nonporous structures that can build high capacitance. To build a textile capacitor, nanomaterials such as CDC or conductive carbon powder are impregnated into yarn by screen printing. Carbon-painted textiles that are conductive transfer electricity whereas the CDC electrodes are separated by a solid electrolyte. Put together, this can look like a tiny patch on a garment but would hardly measure enough capacitance to release a 50,000-volt shock. Perhaps it may be able to light up a blue LED stripe for aesthetic value.

GAMORA'S SWORD, GODSLAYER

WHEN: Guardians of the Galaxy, Guardians of the Galaxy Vol. 2, Avengers: Infinity War	
WHO: Gamora	
SCIENCE CONCEPTS: Physics, nanotechnology, metallurgy	

INTRODUCTION

If you were to tear this page out of this book, scrunch it into a paper ball, and throw it at your index finger, you would barely feel it. However, imagine running that same finger quickly along the edge of this page. You would likely get a paper cut that would be deep enough to hurt but probably not deep enough to bleed. What are the physics of a cut? For centuries we have made tools and weapons that have optimized the variables of a cut and sharpening materials to give you the closest shave, a perfect slice of cheese, or an enemy's severed head. Here we will learn how to make the sharpest blade in the universe: Gamora's Godslayer.

BACKSTORY

In *Guardians of the Galaxy*, each team member has his or her signature weapon. Groot has his extending vines, Star-Lord has his dual blasters, Drax has his twin daggers, and Gamora has her sword, Godslayer. This collapsible sword, when drawn, extends double blades and conceals a dagger in its hilt for close quarters combat or to throw. The first time we see Gamora use this sword she effortlessly slices off Groot's arms and drives the blade into his gut. In *Guardians of the Galaxy Vol. 2*, she uses Godslayer to deliver the final blow to the Abilisk with a downward strike, slicing its front in half from the throat downward. Her sword possesses an energy core in its hilt that makes the sword lighter despite being made from very dense materials.

THE SCIENCE OF MARVEL

Gamora's Godslayer is very effective at slicing her enemies with a swing or stabbing them with a thrust. In order to understand how her sword has been optimized for both these actions, we need to consider the force of her sword. As you would imagine, the greater the pressure exerted on an object, the more damage dealt. If you picture the Godslayer's shape, the parts of the sword that are the most lethal are the edge and the point of the blade. At both of these parts of the sword, surface area is reduced significantly. If Gamora thrusts with the sword's point, the reduced surface area of the very thin blade delivers significantly more pressure than would a slap with the flat of the sword. However, there's more to it than just the force of one object hitting another with low surface area.

Remember Gamora's first encounter with Groot—she sliced both of his arms off with each swipe of Godslayer. If you were to slow down the action and zoom in to the blade cutting into Groot's woody skin, you would see the blade acting like a wedge in a fracture in the wood. This wedge would cause a force at the point of contact that would separate the intermolecular forces that hold together the fibers of the arm. At about a third across the length of the sword, the wedge has pushed through the fracture and traveled across a cross-section of his arm. The forces holding together Groot's arm are probably the interactions between uncharged molecules that occasionally form dipole moments that create some force of attraction.

While surface area is an important variable to enhance the cut, there is also the applied force of Gamora's swing that improves its ability to cut. One way Gamora increases this force is by continuously swinging Godslayer while maneuvering around her enemies. The arc of this continuous motion increases with the acceleration of the swing and the mass of her sword. So a heavier sword accelerating faster would supply the most force that can be applied over a small surface area. However, there are limits to Gamora's strength, and the way she handles Godslayer makes the weapon seem as light as a feather. One way this can be achieved is through the action of the energy core located in the hilt of this sword. We can speculate that in this hilt is a spinning gyroscope; it is rotating clockwise so quickly that it is creating angular momentum along the axis of the blade. While it doesn't necessarily make the sword lighter, it can offset balancing forces along the blade, making the Godslayer, which would normally be too heavy for Gamora to maneuver, handle like a much lighter sword. This would also explain her sword-wielding style since it would be easier to carry out circular motions that pivot in the range of this gyroscopic force.

THE SCIENCE OF REAL LIFE

Obsidian, a volcanic glass made mostly of silicon dioxide (SiO_2), is one of the sharpest materials in the world. It derives its sharpness from its brittleness and how it breaks to create conchoidal fractures. These fractures break off the parent glass as curved pieces that can have an incredibly sharp edge. The intersecting point, or edge, of these curved pieces can measure as small as 3 nanometers in width. This is the closest we can get to a monomolecular blade using natural materials, and it was first used as a weapon during the Stone Age!

OBSIDIAN

Obsidian was a key component of the macuahuitl, a wooden club embedded with prismatic pieces of obsidian used by the Aztec, Mayans, Mixtec, and Toltec. Accounts by Spanish conquistadors even claimed that the macuahuitl could cleave the head off a horse in a single blow. These days, some doctors use obsidian blades instead of steel ones since they provide a cleaner cut that heals faster with less scarring. However, obsidian is about as brittle as glass and can shatter very easily despite its sharpness.

The strongest blades in history have often been forged from variations of steel alloys that have achieved a balance between hardness, toughness, and strength. Each of these traits determines how a material will break from abrasive or frictional stress (hardness), shatter when a force is applied (toughness), and deform before breaking (strength). Blacksmiths have optimized these variables by hardening, tempering, and quenching a sword during its forging. For example, the cutting edge of a sword (the top two thirds) doesn't need to be as tough as the bottom third, which will receive blows from an enemy. Toughening the metal of the cutting edge can be done by quenching and cooling, allowing a blade to hold its edge much longer. However, too much toughening can make the blade brittle and easy to break, so tempering and slow cooling the lower third can restore the strength to the blade.

CHAPTER 6: AGGRESSIVE ASSAULTS

Sometimes there are other structural characteristics to a metal alloy. Specifically, Damascus steel has made its mark on history; it was one of the most malleable and toughest materials and was used for swords from the third to seventeenth centuries. Swords forged from this steel were resistant to shattering, could be sharpened to a very fine edge, and had a mottled, flowing surface texture. Unfortunately, the mystery of this forging process was lost during the seventeenth century but it may have had something to do with the ingots of wootz steel imported from India and Sri Lanka. These had trace impurities of tungsten and vanadium, which strengthened the metal. In a 2006 study, Dr. Marianne Reibold and her team at the University of Dresden examined a Damascus blade and determined that its blacksmiths may have engineered the world's first synthetic nanomaterials. Two microstructures were found to imbue these blades with their strength: carbon nanotube (CNT) and cementite nanowires (CNW). CNTs are made up of a hexagonal grid of carbon atoms that can be rolled into a cylindrical tube and make the world's strongest materials, with a tensile strength 100 times greater than steel.

STAR-LORD'S GRAVITY MINE

WHEN: Guardians of the Galaxy, Avengers: Infinity War
WHO: Star-Lord
SCIENCE CONCEPTS: Gravity, tractor beams, gravitational waves

INTRODUCTION
If you were to climb on top of a table and hop off, you would fall and accelerate toward Earth's core at 9.8 meters/second^2. If you were to make a similar climb and leap that same distance in a pressurized space station on the planet Mars, you would accelerate toward Mars's core at 3.711 meters/second^2. Why the difference? Our attraction to

the center of the Earth has to do with several factors such as the mass of Earth and its relative position to our sun (and to a far lesser extent everything else in the universe). In parts of our universe where matter can become very dense or where an object's mass is incredibly high, gravity can even curve the fabric of space-time. What controls these forces? Can we measure them or control them?

BACKSTORY

Star-Lord has several gadgets that come in handy during his adventures with the Guardians of the Galaxy. In his arsenal, he has a gravity mine that snaps open to latch onto the ground and exerts a tremendous gravitational force on either a specific object or in all directions. During his trip to Morag to retrieve the Orb of Power, he uses a gravity mine to pull the Orb of Power from its stasis field and again during his escape from Korath and Ronan's Sakaarian forces. Lastly, he uses the gravity mine to restrain Thanos on Titan momentarily before Mantis subdued him using her empath abilities. So how exactly is Star-Lord's gravity mine working, and is it just gravity or something else?

THE SCIENCE OF MARVEL

If the so-called gravity mine is creating a focused amount of gravity, we should be able to calculate some physical variables of the device. For example, in *Avengers: Infinity War*, the gravity mine is capable of restraining Thanos's arm for a few seconds before Mantis sedates him with her powers. If this device is generating gravitational waves it wouldn't discriminate the way it's shown to when it removes the Orb of Power from stasis or in the way it targeted Thanos's hand. We would more likely expect the incident on Morag, where all of Korath and Ronan's Sakaarian shock troopers are pulled toward the mine as Peter Quill makes his escape. In this regard, the gravity mine seems to be working like a tractor beam using Bessel beams.

A Bessel beam is defined by the concentric circles of light that form around a dark core—basically a doughnut beam. This beam passes through an axicon lens (think of a rigid pentagon-shaped lens that is almost convex) using acoustic, photonic, or gravitational waves to generate concentric circles. They will not diffract or becoming disrupted if an object partly blocks their path. In these cases, the surrounding part of the concentric circle can "heal" the radiation disrupted by the blockage. Because of this, these concentric lasers can actually form on the other side of Thanos and can be used to exert a force as he is being pulled toward the mine. This may even explain some of the energy discharge that is seen when the mine is attempting to pin down Thanos but was not seen during Quill's scuffle on Morag in *Guardians of the Galaxy*.

THE SCIENCE OF REAL LIFE

Bessel beams would work like tractor beams if Thanos was the size of a human cell. Currently, the idea of using Bessel beams falls within the technologies that are used with optical tweezers (see Manipulating Reality). Also, a true Bessel beam with infinite concentric circles would be impossible to make since it would require infinite energy. This hasn't stopped researchers from using Bessel beams with up to eleven concentric circles to trap cells or other microscopic objects. Dr. David Grier of New York University was able to show this on silica beads (not quite the 985-pound Thanos) by changing the phase of the Bessel beam and causing a push or pull on the bead. This idea has also been used with acoustic waves by Dr. Noé Jiménez at the Polytechnic University of Catalonia in Spain; in his work, acoustic waves pass through a spiral grating to generate a vortex of sound that can trap a small object.

However, these approaches do not change gravity as much as they use another physical force to trap an object in a linear path. What about generating gravity through gravitational waves? Gravity was first described by Newton as a permanent and instant force that exists across all objects in space. This theory held up pretty well for hundreds of years until Einstein came up with his theory of general relativity. His work postulated that the fastest thing in the universe is light, thus calling into question the instantaneous character of gravitational forces around the universe. This guided Einstein to rework his understanding of gravity to consider that space-time is curved around objects with more mass and that this curvature generates gravity. He even theorized the existence of gravitational waves, which decades later were confirmed by experiment.

While theorized by Einstein, the problem with measuring gravitational waves is that they are represented as strains in our reality. This is particularly difficult to measure since if you wanted to use a ruler to measure that wiggle in time and space, the ruler itself would fluctuate with the change you are trying to measure. This is what led to the building of two Laser Interferometer Gravitational-Wave Observatories (LIGOs) in the United States. These identical L-shaped facilities stretch with two perpendicular arms that reach 4 kilometers away from their intersecting point. These facilities project a high-powered (100 kW) laser through a beam splitter into these perpendicular 4-km arms only to reflect this beam back where it can interfere with itself. If a LIGO facility experiences a gravitational wave, this interference pattern will detect a phase change.

Here's the crazy thing: when the LIGO detected its first gravitational waves (from a binary system collapsing into itself), it had to measure a wiggle occurring in the range of 10^{-18} meters. This is equivalent to measuring the distance to the nearest star, with an accuracy limited to the width of a human hair. To carry this out, the Livingston, Louisiana, and Hanford, Washington, LIGOs had to have heated vacuum-sealed laser chambers operating in unison. Together they were able to measure gravitational waves from two black holes merging that occurred billions of years ago. With additional LIGOs being built, we will soon be able to coordinate the triangulation and spatial positioning of these events using gravitational waves.

Chapter 7

Mechanical Marvels

ARTIFICIAL INTELLIGENCE

WHEN: *Iron Man, Iron Man 2, Iron Man 3, The Avengers, Avengers: Age of Ultron, Avengers: Infinity War, Spider-Man: Homecoming*

WHO: J.A.R.V.I.S., Ultron, F.R.I.D.A.Y., Karen, Vision

SCIENCE CONCEPTS: Neural networks, artificial intelligence, systems neuroscience, learning

INTRODUCTION

The first electric programmable computer, the Colossus, was built in 1943 and used to do only one thing: crack encrypted messages sent from German forces during World War II. It was made from 1,700 vacuum tubes that took up the size of a living room and could carry out Boolean logic operations using paper tape as an input. Today, we have computers that fit in our phones and can recognize our voice, play our favorite music, and map a drive into a new neighborhood. More importantly, we have succeeded in moving complex operations away from living room–sized computers and delegating them to decentralized networks of servers. The technological innovations continue unabated, with artificial intelligence (AI) presenting the current focus of much interest. What is the future of computing, and what advances are we making in creating AI? Should we fear the creation of a megalomaniacal robot bent on the extinction of the human race?

BACKSTORY

Tony Stark (Iron Man), James Rhodes (War Machine), and Peter Parker (Spider-Man) are always assisted by multiple natural language user interfaces in their mechanical suits. These assistive programs provide relevant information regarding the status of the suit, the wearer's health, or the output of various calculations to assess the success of a certain maneuver or attack. During the events of

Avengers: Age of Ultron, Tony Stark and Bruce Banner (Hulk) use the Mind Stone to upgrade one of these programs to oversee the function of Stark's peacekeeping Iron Legion. Inadvertently, they create a rogue AI, Ultron, that quickly concludes that the most effective path to peace is via human extinction. To combat Ultron, the Avengers attempt to use the remaining digital framework of J.A.R.V.I.S. that was modified by the Mind Stone to create an AI more empathic to the human condition. The result is Vision.

THE SCIENCE OF MARVEL

In Bruce Banner and Tony Stark's first peek into the Mind Stone's power, they identify what appears to be a new coding language in the Infinity Stone. Through the technology available at Stark Industries, they visualize this language as a holographic projection Banner describes as "neurons firing." It would seem that the blueprint provided could give sentience to a computerized brain, a capacity that does not exist within the realm of current human-defined programming. Current forms of AI are tailored to solve a human-defined problem and are incapable of defining problems themselves through observation. For example, human programmers can collect satellite data, overlay it with city infrastructure parameters, develop pathfinding algorithms, and then make an app that can find you a restaurant down the block. There is a tremendous amount of human rule-building in this example. While there remain several hurdles in developing artificial intelligence, this is probably the biggest obstacle to machine sentience.

During Ultron's first moments he is given a single user-defined problem from Tony Stark in the words "Peace in our time." Imbued with power from the Mind Stone, Ultron proceeds to immediately redefine these words—the only solution being the extinction of humans. At this point he isn't thinking like a computer; he is behaving like a transcendent megalomaniac mind walled within the confines of a computer. Unlike a human, however, Ultron gains knowledge by flickering through words and pictures found on the

Internet despite having no sensory organs. This iterative process probably relies on deep neural network learning algorithms that allow him to perceive the world around him. Neural networks are partly modeled on the activity of neurons in the brain, which receive information from many other neurons, process it, and relay it to another neuron. In silico, this process involves taking information from many sources into one layer of processing, then relaying it across to a second layer of processing, then a third layer, etc.

To oversimplify this, Ultron sees a photo of Tony Stark and uses many identifiers to classify aspects of his appearance before he concludes he is, in fact, looking at Tony Stark. For example, one layer of processing may decide if the face has facial hair, another layer will establish if the facial hair is a goatee, etc. Through the process of attrition, each layer of the neural network gives Ultron a more and more complete interpretation of the picture he is seeing. This information about a single face still needs to be abstracted and built on an even larger interpretation of what a human face is in the first place. Within this scope, what is an edge of a photo? What is a corner? This leads Ultron to train his computational cognition on millions of photos of human faces without facial hair, with facial hair, at different ages, of different ethnicities, etc. The very fact that Ultron goes on to recognize the world, its people, and its history within fifteen seconds of coming into existence is a testament to the power imbued to him through the Mind Stone. In essence Ultron was given a humanlike computer brain capable of awareness like a human. He perceives his environment through iterative analytical deep learning. He can quickly apply this thinking to problems his sentient mind can define and discover more nuanced information in large data sets.

THE SCIENCE OF REAL LIFE

The likelihood of a rogue AI attempting to destroy the human race seems pretty distant at this moment. However, we have made impressive advances in simulating parts of a living brain in silico. In an ongoing international collaboration, a Swiss group is leading the digital reverse engineering of brain microcircuitry. This project, called Blue Brain, aims to use neuroscience research in animal models to simulate the function of a neocortex through its types of cells, connections, firing patterns, and resulting circuits. In the first step of this project, the scientists recreated a single neocortical column in the rat brain with ten thousand neurons and 10^8 synaptic connections among them. This part of the brain is supposedly linked to higher cognitive functions, such as conscious thought. This feat required the processing power of the Blue Gene and Magerit supercomputer clusters, and the project has since scaled up this neocortical column to a mesocerebral circuit consisting of one hundred neocortical columns.

Blue Brain does not provide a complete picture of rodent brain function. It lacks several layers of the biological function such as gene expression, the role of the space, cardiovascular systems, and the metabolic needs of the brain. It is expected that these layers of organization as well as several other cell types in the brain will be integrated in future iterations of the Blue Brain. Also, a rodent brain simulation gives insight into the conscious thought of a rat, and it lacks the organization to make inferences about human brain function. A rat neocortical column with ten thousand neurons has only been scaled up to a model of one hundred columns, whereas the human neocortex consists of one million columns, each made of sixty thousand neurons. This hasn't stopped the Swiss researchers from beginning to model the function of the human brain. As of 2018 it is in its early stages of development.

Perhaps the most significant advances in machine learning have been made by Google's DeepMind research group. DeepMind developed AlphaGo, the first computer capable of defeating the world's best Go player. Losing to a computer AI isn't new. However, when you begin to consider the 2.08×10^{170} different moves that can occur in a game, it becomes a daunting challenge for typical AI to bash through all the possible moves. Because of this, the engineers at DeepMind developed an AI that could learn from and intuit its moves by observing thousands of other games. These other games from a training set can be used to make a smarter Go AI, which can be trained against older versions of itself over and over again. Interestingly, this neural network–reinforced learning algorithm can be general purpose, meaning any kind of visual information is enough for this AI to learn. It is currently being trained on video games from the seventies and eighties and is slowly being introduced to larger problems. For example, DeepMind has been able to use variations of its software to put on a virtual pair of legs in a three-dimensional environment and teach itself to walk.

CYBERNETIC PROSTHETICS

WHEN: *Captain America: The Winter Soldier, Captain America: Civil War, Guardians of the Galaxy, Guardians of the Galaxy Vol. 2, Avengers: Infinity War, Luke Cage*

WHO: Winter Soldier, Nebula, Misty Knight

SCIENCE CONCEPTS: Brain-computer interfaces, neuroscience, neuroplasticity, cybernetics

INTRODUCTION

Our brain has been an incredible asset during human evolution. It is one of the driving forces behind our success as a species and it has been honed to help us perceive, interact, and shape the world around us. Specifically, our brain and peripheral nervous system are good at adapting to change. For example, when we lose a limb, not only does our gross anatomy change, but the way our brain represents the information that used to be processed by that limb changes. While prosthetics that leverage a neural link to the brain remain in their infancy, basic research has begun to uncover the mechanistic underpinnings of brain-computer interfaces that may influence the next generation of prosthetics.

BACKSTORY

Throughout the MCU, many characters have cybernetic enhancements. When Bucky Barnes and Misty Knight each lose an arm, they undergo surgery to connect a metallic arm. Characters from *Guardians of the Galaxy* have moderate cybernetic enhancements, such as those seen on Rocket's back, or major cybernetic prostheses and implants, such as a majority of Nebula's body. These enhancements give their user an edge in combat, with materials that are more resilient than mere bones and flesh; they are also replaceable or repairable when damaged. The use of a mechanical arm also seems to allow users to circumvent pain if it will impede their abilities in the field. This is seen in almost every interaction Nebula has with the Guardians of the Galaxy; she cuts off various parts of her body to get out of many difficult situations.

THE SCIENCE OF MARVEL

Here we'll focus on the "hardware" of one system—a bionic arm—and look at how it could be repurposed to broadly enhance many other organ systems or limbs. While exceptions always exist, most cybernetic limbs need to communicate with the central nervous system and with the bidirectional traffic of information. If a bionic arm needs to pick up an apple, your brain needs to be able to send a coded signal to that limb. Once the arm has picked up the apple, it needs to send a signal back to the brain giving such data as the weight of the apple and its texture.

Consider how a normal human arm works. When your brain decides to pick up a book like this one, it sends a descending electrical signal along the axons of your spinal cord that takes a turn through several neuronal fibers at cervical vertebrae. Those neurons then transmit an electric current across various muscle groups along your arm and across the two hundred thousand neurons in your hand to articulate a gesture that allows you to pick up a book. Once the book is in your hand, a returning pattern of neuronal firing gives your brain information about the book's weight, texture, and density. Cutting that flow of information at any point requires a great deal of rewiring and training. In the case of Bucky Barnes, after he loses his arm, the nerves from his brachial plexus were probably threaded into a peripheral nerve interface (PNI). This small piece of technology would transduce electrical information received from the brain into a machine processor that would effectuate motor responses in a mechanical arm. The PNI itself could take on forms such as a helical platinum ribbon that can wrap around the nerve or a long strip that is sutured to the tissue surrounding the nerve.

Other forms of cybernetic arms can also receive information related to muscle contraction activity from electrodes on the surface of the skin. In the *Luke Cage* TV series, detective Mercedes "Misty" Knight, who lost her arm to the blade of a katana in *The Defenders*, has it replaced with a new arm from Rand Enterprises. This arm is connected below the shoulder with a small wire connecting an electrode to her upper arm. The sensor reads electromyographic (EMG) signals in order to provide the mechanical arm with user input. Electromyography relies on sensing the electric potential of a typical muscle when it contracts. This potential is created every time a group of muscles contracts and can be decoded by the prosthetic arm.

THE CASE OF NEBULA

In the case of Nebula from *Guardians of the Galaxy*, it appears that her limbs are more "optional," suggesting that most of her body is mechanical save for her brain and spinal cord. This is most evident in a scene from *Avengers: Infinity War*, when Thanos stretches out Nebula's entire body, revealing its mechanical parts. In such a case, very little of her nervous system innervates the extremities of her body. Nebula probably uses many PNIs that don't plug into the proximate areas of a limb that has been lost; rather, they're placed throughout what is left of her spinal cord. These implants could use synthetic neuromorphic neurons that would allow her to not only interact with her surrounding world through a mechanical limb but also sense information through her limb.

CHAPTER 7: MECHANICAL MARVELS

THE SCIENCE OF REAL LIFE

The science explained earlier isn't too far in the future compared to what we have today. In fact, many of these technologies were originally developed as medical diagnostic tools to measure or modulate the health of various tissues. For example, EMGs are commonly used to measure the function of muscle and detect any neuromuscular abnormalities. Similarly, electrical pacemakers regulate heart contractions, and vagus nerve stimulators help control seizures. It's not surprising that much of this technology has been exploited in engineering a human prosthetic arm that can function as a normal arm. However, there are some difficult hurdles to overcome before we can see an amputee use his or her prosthetic arm with the dexterity of the Winter Soldier.

Using various implants to regulate a mechanical limb is feasible. However, learning to use such a system correctly can be incredibly difficult and time-consuming. Consider how an artificial limb has its own internal wiring and components; your brain, on the other hand, uses a completely different substrate to communicate with its tissues. It's like putting two people in a room who need to work with one another but speak two completely different languages. With a lot of time and training, they can learn to speak with one another (albeit with a bit of an accent). In order to use these types of prosthetics, a patient's nervous system and brain literally need to "learn" the arm and how it functions. In this regard, patients may need to spend hundreds of hours carrying out preprogrammed motions in the prosthetic arm and syncing them to the thought of moving a limb and repeating this over and over. This process may involve, for example, activating a prosthetic arm to make a fist and then asking the person using the arm to mentally form a fist while recording the electrical potential under her skin using EMG electrodes or PNIs inside her tissues. The use of EMG to effectuate control over a prosthetic arm is rather limited, however. For example, only a handful of motions can be carried out with the contractions of muscles in an arm. When you consider the range of motion of an arm and the error rate of these new technologies, it's easy to see what a huge task it would be to train one's mind to properly use a bionic arm with the freedom of a normal one.

Currently, advances in prosthetic technology are innovating in a variety of ways. Claudia Mitchell, one of the first people to be outfitted with a bionic arm, had her arm designed by the Rehabilitation Institute of Chicago. After losing her arm in a motorcycle accident, she had surgeries to "rewire" the nervous tissue of her arm into the muscles of her chest. As a result of the surgeries, these nerve bundles were stimulating actual muscle tissue as opposed to ending at a limb that didn't exist anymore. Now, the mental thought of making a fist causes a series of twitches over her chest that map out the typical firing patterns for a hand, only on the chest where they can be detected by several EMG electrodes. These EMG electrodes reroute information from these signals into the mechanical arm that she controls.

SPIDER GRIP AND WALL CRAWLING

WHEN: *Spider-Man: Homecoming, Captain America: Civil War*

WHO: Spider-Man

SCIENCE CONCEPTS: Spider morphology, biomechanics, mechanical engineering, biomimetics

INTRODUCTION

In the animal kingdom, most creatures that chose to live on land were subjected to the constraints of gravity. However, there are always exceptions: spiders and multiple species of geckos evolved the ability to stick their feet onto almost any surface. This ability to modify the friction of their limbs allows these animals to sustain multiples of their own weight and rapidly scale various low-friction surfaces. What happens at the microscopic level between the surface of a spider's legs and a glass window that allows them to stick? Can we translate this ability to tools for human use or Spider-Man's wall-crawling powers?

BACKSTORY

Within Spider-Man's repertoire of talents is his unique ability to stick and release his grip onto any surface with just the tips of his fingers and toes. In addition to sustaining his own weight, his grip can withstand the force of his powerful leaps, hold his balance on the edge of a freighter truck, and allow for a 170-meter climb up the Washington Monument to be completed in under a minute. Impressively, his grip adheres to any surface regardless of friction (concrete to glass) or at any angle relative to gravity. His grip can even be used as a grappling maneuver as seen in *Spider-Man: Homecoming* when he kicks an ATM robber, sticks him to his foot, and throws him across the room. So what happens at the microscopic level when Peter Parker's hand touches concrete?

THE SCIENCE OF MARVEL

If we want to understand how Spider-Man's adhesive grip works we first need to consider how a real spider's grip works. Spiders already have a unique gait that balances their weight across eight points of contact; that doesn't directly apply to Peter Parker's abilities. However, it does correspond to the surface contact he makes. He has at least ten points of contact between both hands and two larger contacts corresponding to the balls of his feet. If you were to look at the ends of a spider's legs under a microscope, you would see small structures called scopulae that look like tiny little brushes. If you were to zoom in even further you would see that these brush fibers form tiny little triangles at their ends. These microscopic structures form the basis of Spider-Man's incredible grip and allow him to make intermolecular interactions with his climbing surfaces.

These sticking forces rely heavily on van der Waals forces that hold most materials together. When Peter Parker's index finger digs into a piece of concrete, these microscopic hairs wedge in between molecules and anchor themselves with these little triangular hitches. Once inside the concrete, these microscopic hairs create dipole interactions within his climbing substrate, functioning like an intermolecular Velcro. Typically, these are considered relatively weak forces, but the density of these hairs digging into a climbing surface collectively contribute to the grip's overall strength.

How does Spider-Man unstick his hand once he's established his grip? In this regard, his ability may be similar to that of a gecko, since a spider has more points of contact to apply its scopulae at varying angles. While still working through the same creation of van der Waals intermolecular forces, a gecko's digits have the hair-like structures set at oblique angles, which provide a greater surface area for sticking and are capable of supporting more weight. In addition to this, these hairs (called setae in geckos) are very flexible, allowing them to absorb the force of their springing step, which allows a gecko to travel much quicker than a spider does.

Peter Parker's super strength contributes to his finger strength and climbing ability when he doesn't have to worry about scaling frictionless surfaces such as glass. Parker probably prefers to use ledges and grooves in his climbing surface to gain more upward momentum. Human fingers don't actually have muscles and rely on tendons extending from the palm and forearm to articulate motion. When Parker is grabbing onto a ledge exerting a grabbing force he's probably exerting work on his flexor muscles that extend from the bottom of his forearm and palm. In contrast, when letting go of his grip he's exerting work on his extensor muscles on the top of his forearm and hand. While similar to human muscle groups, chances are the increased tensile strength of Parker's tendons allows him to carry more than his own body weight. Spider-Man's wall crawling is the combined effects of his ability to dig his hands into the intermolecular space of any given surface paired with a biomechanical advantage!

THE SCIENCE OF REAL LIFE

Spiders are masters of scaling frictionless surfaces. The jumping spider *Evarcha arcuata* typically uses its tarsal claws to scale normal surfaces, but on smooth surfaces it uses its tufted claw, the scopula.

SPIDER STRENGTH

Dr. Antonia Kesel at the University of Bremen in Germany found that the average spider has 624,000 setules that cover a point of contact area of 1.7×10^5 square nanometers. Looking at the adhesive force of a single setule, Kesel showed that it can hold the force of 38.12 nanonewtons. While this may seem small, relative to the weight on one of these spiders, if this spider were to lay down all of its appendages it could likely hold up to 170 times its weight.

Geckos have analogous hair-like structures called setae that split up into spatula-shaped tips that not only stick due to van der Waals forces but also use electrostatic forces. A study led by Dr. Hadi Izadi at the University of Waterloo in Canada discovered that contact electrification (similar to what happens when you drag your feet on a carpet) can create attractive forces between the gecko's padded feet and its contact substrate.

How close are we to scaling walls like Spider-Man? It depends; if you were to ask University of Cambridge professor David Labonte, he'd tell you that a gecko is probably the largest an animal can get and still have the sticky adhesive pads carry its weight. In a study he published, Labonte examined 225 climbing species to identify allometric trends between a given species size and its footpad area for climbing. Upon identifying a 200-fold increase in surface area of footpads between a mite and a gecko, it appears that a human would require an unrealistic 40 percent of its body be covered with scopulae to hold the grip that Spider-Man shows. This suggests that the gecko may be the largest animal to evolve this mechanism of adhesion for climbing. However, this is a constraint on evolution, not human ingenuity. Elliot Hawkes, in work conducted at Stanford University, designed a weight-distributed biomimetic gecko glove that allowed him to scale a glass wall at Stanford (albeit, not at the rate of a gecko).

Many of these studies examining the basic physical forces that allow for adhesion are particularly useful for human technologies. For example, a biomimetic design that incorporated the mechanical properties of a gecko's sticky pads is being used for new surgical bandages. Dr. Alborz Mahdavi of MIT carried out research that developed a biodegradable polymer with the modified topography mimicking a gecko's footpad; it may one day replace sutures. Similarly, such technology has formed the basis of new super adhesives such as Geckskin that have been used for DARPA's Z-Man program to enable soldiers to scale vertical surfaces. In a 2012 demonstration of Geckskin, a 16-square-inch patch on a vertical glass wall could hold a static load of 660 pounds.

MANIPULATING REALITY

WHEN: *Thor, Thor: The Dark World, Thor: Ragnarok, Avengers: Infinity War*

WHO: Loki, Frigga, Thanos

SCIENCE CONCEPTS: Photonics, optical trapping, lasers

INTRODUCTION

Light is one of the most important physical forces in our world. Our perceptions of light and darkness can tell us how close or far an object is, its shape, and its position relative to other objects. The wavelength of visible light can add an optical texture to objects that allows us to infer the ripeness of a fruit or whether an animal presents a threat. However, our interpretation of light isn't restricted to the visible spectrum. The higher frequencies of light can be used to cook our food in microwave ovens and broadcast radio and TV programs. Light, or electromagnetic radiation, is an important feature that shapes our reality. Is it also possible for us to manipulate light like Loki or the Reality Stone can?

BACKSTORY

In the MCU, both Loki and the Reality Stone appear to be able to shape light to create tangible illusions. Loki uses his powers to transform himself into Odin (*Thor: The Dark World*), make his armor appear, or even make copies of himself (*The Avengers*). Each time Loki uses these powers, a shimmering light appears to conceal the illusion in question. These light constructs also have the ability to take form in the physical world as actual matter. This phenomenon is more pronounced when Thanos uses the Reality Stone to exert his will over reality. However, this power is only temporary. This is seen when he turns Drax and Mantis into a pile of bricks and ribbons.

THE SCIENCE OF MARVEL

Loki's ability to manipulate light appears to have been taught to him by his adoptive mother, Frigga, who is the only other Asgardian in the MCU who appears to have this inherent ability. Knowing that Asgardians have mastered technology so it often seems magical to humans, we can speculate how these constructs can be made through our understanding of light. If we can assume that Loki and the use of the Reality Stone exert a certain mastery over electromagnetic radiation, how would that shape our perception of reality?

Focused light can exert a force on an object that allows it to be used like a microscopic tractor beam. If you were to visualize all the paths of light traveling out of a lens you would see a three-dimensional cone of light converging at a single point before diverging. The photons of light exert force within that cone that can push micrometer-sized particles toward the convergence of the lens's light path. After a particle reaches the convergence, it continues along this path. However, if you add a second light source coming through a lens from another angle, they can simultaneously converge on that particle, allowing it to be fixed in three-dimensional space. If Loki has influence on light that can move small particles of dust, he can use light from many angles to assemble a "skin" that could have light of other wavelengths projected onto it like a dynamic projector screen. For subtlety he could even use the nonvisible light spectrum to move particles, and the visible light spectrum to project colors onto his dynamic dust particle skin. This in essence would create a volumetric representation of light, using an image that would be as resolved as the smallest particles he can find in the atmosphere. This would explain that extra white light that shimmers whenever his illusions fade or appear since they are concentrated points of light that are losing their microscopic particles from a converging point of light.

LIGHT AND FORCE

Light has the properties of both a wave and a particle, and due to its ability to behave like a particle, it can generate force. This is why a comet has two tails, since solar radiation applies pressure on the dust tail to form a second radiating arc away from the sun.

While his light projections may in fact be hiding objects Loki has in his possession, it sometimes appears as if he can take one of these illusions and turn it into a tangible object. In this instance, he may be turning light into matter through a rearranged general relativity equation of $E = mc^2$. He could use high-energy photons

CHAPTER 7: MECHANICAL MARVELS

colliding with one another but this might require too high a level of energy. Similarly, if Loki has access to that kind of power, he probably wouldn't be using it to create matter; he could just zap people.

THE SCIENCE OF REAL LIFE

The reality of optical trapping is much older than you might imagine. And today, advances in technology and innovations in the uses of lasers have brought us closer to manipulating the visible spectrum in ways similar to Loki's illusions. The idea of light exerting a force goes back as far as 1619 when Keppler described the two tails of a comet's trailing path, and the groundwork for optical trapping was begun in the 1970s and 1980s by Dr. Arthur Ashkin at Bell Laboratories. (In 2018, Ashkin, then age ninety-six, shared the Nobel Prize for Physics for his work related to optical trapping.) Ashkin's work makes it possible to move living bacteria, manipulate the organelles inside a cell, and sort living from dead cells in a liquid medium. In fact, optical trapping has become a vitally important tool for manipulating cells and exploring the forces that move subcellular objects in the cell.

The ability to use these forces of light to create a volumetric display is a little more recent and was pioneered by Dr. Daniel Smalley at Brigham Young University. Inspired by *Star Wars: A New Hope* and the famous "You're my only hope" R2-D2 hologram, Smalley's group wanted to create a volumetric display that could be observed in 360 degrees. In his display, optical traps were used to trap a cellulose particle in "thin air," where the illumination of the laser trapping the particle imbues it with a glow. This particle can then be dragged in space, and due to a phenomenon known as optical persistence, an observer perceives a line in space. (Optical persistence is the perception of a visual stimulus for a little longer than it exists, allowing for shapes and forms to be traced or rastered incredibly fast.) Currently this process requires a sophisticated setup of lasers and mirrors and the size of the volumetric display is limited to about 10 cubic millimeters.

What about the creation of matter from light? The earliest postulation of this theory was made in the 1930s by Gregory Breit and John A. Wheeler; they hypothesized that if two photons collided with one another, they could create a positron and an electron. This may not be the matter you'd expect Loki or the Reality Stone to fashion, but as of 2018 a group at Imperial College London are ready to give it a shot. This research, led by Dr. Steven Rose, involves building a photon collider that will use a powerful gamma beam in a heated chamber to generate electron and positron pairs in a measurable amount.

Independent of turning electromagnetic radiation into light, there is also a unique way in which light can behave like molecules. In a 2013 study, a group at MIT and Harvard were able to create a chamber of supercooled rubidium atoms that, when a weak laser was fired through it, created a particle called a Rydberg polariton. These particles moved significantly slower than the photons entering the chamber and appeared to emerge as pairs that looked like bonded photons. Follow-up experiments showed that triplets of photons could also emerge with a stronger bonding energy between these "photonic atoms."

Chapter 8

Stupendous Scenarios

GAMMA RADIATION

WHEN: *The Incredible Hulk, The Avengers, Avengers: Age of Ultron, Avengers: Infinity War, Thor: Ragnarok*

WHO: Hulk, Abomination

SCIENCE CONCEPTS: DNA structure, mutation, genetics, DNA damage

INTRODUCTION

The DNA that makes up our genome stays with us from the moment we are conceived and lasts until long after we pass away. During your embryogenesis, that one copy of your genome becomes two, then four, and so on. As an adult you've made your way from a single-cell embryo to 37.2 trillion cells with different functions. In fact, over your lifetime, you've divided your cells 10^{16} times. That is a lot of copying of information and there are likely to be a few mistakes along the way. Even then, that information is constantly being attacked by different chemical mutagens and electromagnetic radiation that can change the function of DNA and proteins in dramatic ways. Your cells have ways to recover from this type of damage, but in the MCU your options are limited—it may be a good time to buy a pair of stretchy pants.

BACKSTORY

Bruce Banner had his first transformation into the Hulk in a botched experiment attempting to use gamma radiation on himself. While never intending to make a weapon or Super Soldier, Banner was trying to increase the resilience of cells to radiation. His equipment malfunctioned; every cell in Banner's body transformed and with it a signature of lingering gamma radiation. In *The Incredible Hulk*, Dr. Samuel Sterns deduced that the transformations are initiated from gamma radiation pulses coming from Banner's amygdala. Banner's gamma-irradiated blood is later used to push Emil Blonsky's Super Soldier transformation into the grotesque Abomination. In these circumstances, gamma radiation acts like a catalyst that pushes normal physiology into the realm of nuclear rage.

THE SCIENCE OF MARVEL

Gamma radiation is not just central to Banner's transformation, it was the subject of his life's work. Remember, he was initially recruited by the Avengers to trace the location of the Tesseract because of the low amounts of gamma radiation it emits. His research leading up to his transformation also focused on exposing pulses of gamma radiation on concentrated myosin protein in order to develop radiation resilience in human tissues. These research interests and outcomes raise many questions, such as what is gamma radiation and what does it do to living tissues?

First and foremost, what is gamma radiation? Consider an atom: its neutrons and protons that make up its nucleus, and its orbiting electrons. Most chemical reactions involve the exchange of electrons between different elements to form bonds, while never modifying the number of neutrons or protons present in the nucleus. In contrast, any change of neutrons or protons within that element involves nuclear decay, causing radioactive emissions. This is because a strong nuclear force maintains the integrity of the nucleus that, when disrupted, needs to release a large amount of energy. This nuclear decay results in the release of alpha (a helium nucleus: 2 protons + 2 neutrons), beta (1 electron), and gamma emissions (high-energy photon). This gamma radiation is the highest energy wave on the electromagnetic spectrum, with a wavelength shorter than the size of an atom (10 picometers). This high-energy radiation can pass through atoms and cause other atoms to release electrons when bombarded with gamma radiation. Therein lie its effects and danger.

Gamma emissions are known as ionizing forms of radiation. This means that when they hit an atom, they overload it with so much of their energy that they can force the release of electrons within its atoms. This can result in the generation of reactive oxygen species capable of inflicting oxidative damage on naturally occurring chemical bonds that maintain the integrity of proteins and DNA in Bruce Banner's cells. In the Hulk, this profound damage may contribute to a whole body change, affecting the DNA in each of his cells. Gamma radiation is known to cause double-stranded breaks of the DNA double helix, effectively shattering the entire genome. Following this shattering, repair mechanisms set to work to reassemble Banner's DNA, creating a slew of mutations across his genome. This one-in-a-zillion genomic reassembly resulted in the genetic changes that underlie Hulk's physiology when he transforms. Given this chance occurrence, his new genetic architecture encoded for "ultra-functional" forms of his genes that could have improved function.

One example of genes that might be affected during Banner's transformation to Hulk could be those that regulate muscle metabolism and function. For example, insulin growth factor and growth hormone already initiate muscle growth in humans. Perhaps a reconstituted growth hormone would include known mutations that improve its capacity to bind receptors that signal muscle growth, which would sustain its effects longer than in a normal human form.

THE SCIENCE OF REAL LIFE

The chance reassembly of genetic sequences of 3,200 megabases organized in twenty-three chromosome pairs, in 37.2 trillion different cells with different functions, would require some pretty amazing luck. Even then, this assumes that breaking a gene a certain way will force its reassembly into a superior functioning form. This is analogous to taking many porcelain plates, shaking them inside a container with some Krazy Glue, and expecting a collection of Fabergé eggs. Sadly, the effects of gamma exposure would not have such a positive outcome (yes, in this specific case, turning into a giant green rage monster every now and then is still a better outcome). Incidents with nuclear power plant failure (Fukushima and Chernobyl) and the fallout from bombs dropped during World War II and at field test sites have provided clear documentation for the many effects of radiation exposure. From these tragedies we have a picture of what gamma radiation does to the human body.

Banner's original experiments involved subjecting the protein myosin to gamma radiation to study cellular resilience. When considering ionizing radiation at the levels experienced by Banner, it probably doesn't matter what protein we are studying since the effects of ionizing radiation will affect most proteins in a similar fashion. We can look to the literature and find a handful of studies that examine the effects of gamma radiation on bovine serum albumin (BSA). BSA is purified from water soluble plasma found in cow's blood and it is a go-to source of abundant protein used in various laboratory settings. Within BSA is a chain of amino acids that fold into a series of organized helices and sheets occupying a spherical volume. Imagine a gymnast's ribbon flying in the air, forming structures around the flicks of a wrist to make spirals and weaves that suspend themselves in the air. Exposure to gamma radiation results in the fragmentation, degradation, and crosslinking of BSA's structure, ultimately changing its function. This would be like a pair of scissors cutting the gymnast's ribbon and stitching adjacent weaves together to form unpredictable tangles.

At a larger scale, these pronounced changes would result in a lot of cell death. In reality, Banner would have died instantly or suffered from acute radiation sickness. At the highest intensities of gamma radiation he would have effectively killed the cells in his bone marrow, gastrointestinal tract, and central nervous system. A loss in bone marrow would result in a drop in white and red blood cells, making him easily succumb to infections and bleeding. While this can be partially remedied with bone marrow transplants, the destruction of gastrointestinal cells would keep him from absorbing any nutrients from his food. At its worst, damage done to

Banner's central nervous system would manifest through swelling that would lead to anxiety, confusion, and loss of consciousness within a few hours. In five or six hours these effects would result in tremors, convulsions, coma, and ultimately death.

THANOS'S SNAP

WHEN: *Avengers: Infinity War*

WHO: Thanos

SCIENCE CONCEPTS: Epidemiology, conservation biology, extinction

INTRODUCTION

In the past one hundred years, global population size has quadrupled; by 2020 an estimated 7.8 billion people will be living on Earth. There are many variables that shape human populations and their changes across history. These changes not only affect our culture and civilization but they also have profound rippling effects on our land and the local flora and fauna. Think about it this way: less than 1 percent of all species that ever were are those that are alive today (the rest are lost to time and the fossil record).

What would happen if humankind suffered a catastrophic extinction event? More specifically, what would the next few days, months, and years look like after Thanos's finger snap at the end of *Avengers: Infinity War*?

BACKSTORY

Thanos's motivation to gather the Infinity Stones and harness their power stems from his own traumatic history on the planet Titan. With its ever-increasing population he recognized that his planet would never be able to sustain life if it were to continue to consume its finite resources. Thanos offered a single solution: the genocide of half of his planet's population. Thanos was immediately deemed mad, but in due time his world succumbed to his prediction. Thanos survived and saw that only he could save the universe by finding sentient life and wiping out half of their population indiscriminately.

THE SCIENCE OF MARVEL

During Thanos's first interaction with the human population of Earth, he arrives in the year 2012 with a Chitauri army under the command of Loki. If Loki's invasion had not been stopped by the Avengers, Earth's population of 7,128,176,935 would have been halved. At this point Thanos foresaw an opportune time to secure the Space Stone and kill half of humanity. From his own experiences on Titan he may have realized that Earth's population was reaching a tipping point. He foresaw human overpopulation leading to increased spread of airborne disease in densely populated urban areas and the contamination of remaining freshwater supplies. The overfarming of Earth's resources was sapping the land and causing desertification and the destruction of biodiversity in multiple ecosystems. As these resources neared depletion, the ugliest side of humanity would carry out warfare over remaining minerals, food, and water. Given this outcome, the best thing Thanos could do was murder, using a bejeweled cosmic gauntlet.

What would happen in the first few moments after that fateful snap? We saw a peek at the end of *Avengers: Infinity War*. A 50 percent elimination of Earth's population would collapse the societal infrastructure of Earth's densely populated urban areas, leading to the deaths of millions more. With that 50 percent loss, we lose half of the world's doctors, firefighters, food distributors, and farmers. The loss of this labor and service in densely populated cities would cause a second massive wave of death in hospitals or in areas where food distribution chains would be broken. In rural areas, the effects would be less pronounced and contingent on the remaining survivors being able to harvest the land without any outside aid

(Mennonites would fare well). Unlike with what he did to the people of Gamora's home planet, Thanos was nice enough to turn 50 percent of humanity into just dust; having to deal with three billion corpses would wreak havoc on the remaining population of Earth!

This half extinction would also probably lead to the coextinction of some species and the flourishing of others. For example, several domesticated species of animals that rely on humans to breed would either return to a feral state or die off (pandas, cows, chickens, and small dogs would be the first to go). In contrast, threatened and endangered species might get a second wind and grow in sustainable numbers over the following years. For example, without the same amount of deforestation, hunting, and poaching, several ecosystems in Honduras and Nigeria would recover their indigenous flora and fauna. Perhaps some of the most pronounced effects would occur in the ocean, where the absence of overfishing would allow many marine ecosystems to revive the diminished populations of shark, bluefin tuna, and Atlantic halibut. While a positive outcome at first, this might also cause a cascade of effects that might crash the ecosystem that sustains them. The resulting food chain collapse could lead to the mass extinction of marine organisms. Thanks, Thanos.

THE SCIENCE OF REAL LIFE

When it comes to the motivations behind Thanos's Snap, one book published more than two hundred years ago shows that he was not original in his thinking. In 1798, the economist and cleric Thomas Malthus published *An Essay on the Principle of Population* in which he put forth the thesis that the human population would suffer a cycle of misery. He attributed this to the way in which populations grow geometrically (doubling, tripling, quadrupling, etc.) over time, outpacing food sources that grow arithmetically or incrementally. This suggested that a larger population would lead to an oversupply of labor, reducing wages and creating mass poverty. This philosophy was incredibly influential during the eighteenth century, affecting politics and even inspiring works of Charles Darwin and Alfred Russel Wallace related to the theory of natural selection.

One of Malthus's students adopted these ideas and used their principles to define the terms that contributed to mass starvation in Ireland during the potato famine of 1845–1849. Charles Trevelyan, a British colonial administrator, oversaw the aid efforts occurring in Ireland during this period and is often regarded as a heartless contributor to the deaths of about one million Irish. This was not half of the universe's population, but it did represent 12–18 percent of the entire Irish population. In what may seem like an analog to Thanos's Snap, Trevelyan pulled support for the Irish, believing it to be an example of Malthus's ideas that others could learn from. He forced the Irish to export their oats to England and ended several relief programs that would have sent food to the population during the famine.

THE BLACK DEATH

Another case study in the effects of radical depopulation is the outbreak of bubonic plague that took place in 1347–1750. Dr. Sharon DeWitte, of the University at Albany–State University of New York, showed that there was a silver lining for the survivors of the plague. In demographic data taken from cemeteries pre- and post-epidemic, survivors had higher survival rates in the following years despite recurrent outbreaks of the disease. It should be noted that, unlike Thanos's Snap, the plague probably affected unhealthy populations in Europe and selected for a healthier generation of survivors.

In reality, however, Malthus's arguments, and their conceptual framework relied upon by characters such as Charles Trevelyan, contain many flaws. One particularly important variable that Malthus's arguments did not take into account was human innovation and technology. Specifically, he was blind to many innovations of Europe's agricultural revolution, which was improving the efficiency of farming practices and transforming how food can be produced and distributed. In addition to this, his predictions never held up to how human history actually unfolded. For example, Malthus never would have predicted the advances in farming equipment, food preservation, pesticides, and genetically modified organisms that have allowed us to feed more people than we ever imagined possible. In fact, two hundred years later, fewer people die of starvation and the average human lifespan has doubled.

FORGING STORMBREAKER

WHEN: *Avengers: Infinity War*	
WHO: Eitri, Groot, Thor	
SCIENCE CONCEPTS: Stellar evolution, subatomic physics, chemistry, metallurgy	

INTRODUCTION

Nearly all the life on our planet relies on the radiation given off by our sun. This sun has sat at the center of our solar system for 4.603 billion years, but it won't be there forever. Stars like our sun eventually die out and undergo several changes over trillions of years depending on their mass. While life on Earth has figured out subtle ways to capture the sun's energy and store it within the bonds of sugars through photosynthesis, how does that stellar power look as you get closer to it? How do stars give rise to elements, and what types of materials would require the power of a dying star to forge weapons like Mjolnir, Stormbreaker, or even Thanos's Infinity Gauntlet?

BACKSTORY

Thor relied on the magical Asgardian hammer Mjolnir for 1,500 years. This hammer could call down lightning, create tornadoes, and grant Thor the power of flight. After Mjolnir was destroyed by his sister, Hela, Thor sought out the dwarf Eitri of the star forge Nidavellir to create another weapon, Stormbreaker. Unlike Mjolnir, Stormbreaker has the ability to teleport Thor wherever he needs to go. Both of these hammers, as well as Thanos's Infinity Gauntlet, were made from the Asgardian ore uru, requiring the heat generated from a dying star to smelt the metal so it can be cast into weapons of incredible power.

THE SCIENCE OF MARVEL

The exotic uru metals require a tremendous amount of energy to be melted and poured into a cast. While uru metals are nothing like Earth's metals, we can assume that some of Earth's chemistry may have some transitive properties that can explain the resilience of Mjolnir and Stormbreaker. Consider an atom of iron; since it has a low ionization energy, it is readily able to lose the two electrons in its outer shell to form a positively charged atom. If you were to add more iron atoms to the mix, these lost electrons would form a subatomic glue capable of holding these positively charged iron ions together. This collection of iron ions held together in a sea of delocalized electrons forms the chemical organization of a metallic bond. Because of this bond, metals have very high melting and boiling points. Uru's strength as a metal is likely due to a higher excess of electrons its atoms can lose to become a positively charged atom and a higher number of protons allowing for stronger metallic bonding.

Furthermore, uru is probably used as an alloy, since it is less susceptible to deformation due to the way its atoms are organized. In a pure metal, atoms form a lattice in which each atom is identical in size to its adjacent neighbor. This pure metal can be deformed by any force that pushes these identically sized atoms past one another. The introduction of a differently sized element limits the movement of identically sized atoms across one another. This leads us to believe that the ingots used in *Avengers: Infinity War* may be a combination of different metals, including uru metal, which forms an alloy that can be cast into an Asgardian weapon.

How hot does the forge need to be to smelt these metals together? The dwarves of Nidavellir have spent millennia casting powerful weapons for the Asgardians by harnessing the power of dying stars. This gives us a few options since a star can die a few different ways depending on its mass. Stars exist by maintaining a balance of nuclear fusion radiating outward from their core, with gravity pushing the layers of newly formed elements inward. When a star is ready to die it has used up all the hydrogen fuel in its core, and the pull of gravity is no longer strong enough to keep fusion going. As a result, the outward pressure of fusion causes the star to contract to just over the size of the Earth, or a white dwarf. However, the size of Nidavellir's core is not nearly as large as the size of the Earth and may resemble something closer to a neutron star.

A neutron star forms when an even larger star fuses heavier and heavier atoms in its core (hydrogen into helium, helium into carbon, etc.) until it creates an iron core that cannot fuse any further. This stops the fusion reaction, causing the star to collapse on itself and creating a violent supernova explosion. The resulting effects of gravity pull the remaining atoms in the core of this star so close together that electrons and protons fuse to become neutrons. Nidavellir is about 20 km in diameter but is 500,000 times the mass of Earth, burning at about 1,000,000°F (600,000°C). Assuming the dwarves of Nidavellir have engineered the space station that can contain a neutron star, they may use their steampunk tech to siphon off a fraction of this energy to power their forges.

THE SCIENCE OF REAL LIFE

While not quite an uru ore, the strongest pure metal on Earth is tungsten with 74 protons and up to 6 electrons it can share with adjacent tungsten ions (with a +6 valency in its ionized form). The large number of protons and the electrons it can shed on its outer shell contribute to its strength through metallic bonds. This strength gives tungsten a high melting temperature at 6,191°F (3,422°C). It also happens to be quite brittle in its polycrystalline form (remember Hela shattering Mjolnir into thousands of shards?). However, generating temperatures of 6,191°F (3,422°C) is not as easy for us as it is for the dwarves of Nidavellir.

STANDING NEXT TO A NEUTRON STAR

Assuming we could get close to a dying star, what would it feel like to be as close as Thor was to a neutron star? A neutron star has incredible mass and gravitational forces pulling down so strongly against its surface that electrons are fusing with protons. This force of gravity on the surface of a neutron star is 10^{11} times that of the gravity on Earth. That means that at the distance Thor stood from the core of Nidavellir, he would have succumbed to "spaghettification" (i.e., been stretched *really* thin) by being pulled toward the neutron star before even seeing Nidavellir.

If we wanted to make a tungsten Mjolnir or Stormbreaker, we could use powder metallurgy to take steps toward shaping tungsten in a nonmolten state. This would involve taking tungsten and turning it into particulate powder or mixing in other elemental powders to make the desired alloy. We would then pressurize parts of the hammer (head and handle) in a mold to pack the powder with forces between 10–60 metric tons per square inch using a hydraulic or mechanical press. Once pressed, these parts would be sintered in a furnace where particles of these elements would bond to one another, increasing its density. (It may not involve a dying star but it gets the job done!) Also, just because tungsten is the strongest metal doesn't make it the strongest alloy. In research led by Dr. Xiaochun Li at UCLA, one of the strongest alloys ever made was created by infusing silicon carbide nanoparticles (14 percent) into magnesium. This nanocomposite was created through some clever engineering that was able to evenly distribute silicon carbide nanoparticles into molten magnesium.

POWER STONES AND NUCLEAR FISSION

WHEN: *Guardians of the Galaxy, Avengers: Infinity War*

WHO: Ronan the Accuser, Star-Lord, Gamora, Drax, Rocket, Thanos

SCIENCE CONCEPTS: Nuclear fission

INTRODUCTION

What exactly is power and what happens when you lose control of it? If you consider your smartphone, you may think about the power a battery supplies to the circuits that allow you to use different apps. If the battery were to break and leak its component parts or possibly lose its charge over time, you'd lose power. Now, what would happen if you were to scale the amounts of energy involved to what happens during nuclear fission? Even scarier, what happens if you lost control of nuclear fission? What levels of energy would release? More importantly, can we imagine how such materials could be controlled at the center of the singularity that made the Power Stone ingot?

BACKSTORY

The plot of *Guardians of the Galaxy* revolves around everyone getting their hands on the Orb that stores the Power Stone. The Power Stone is so strong that it can only be wielded by an individual of incredible strength. Those who aren't strong enough, such as the Collector's servant, disintegrate into a purple explosion of uncontrollable energy if they try to grab it. Those who have succeeded in wielding its power often do so using a weapon or garment such as Ronan the Accuser's hammer or Thanos's Infinity Gauntlet. In the case of the Guardians of the Galaxy they had the superpower of friendship! When Thanos acquired the Power Stone he used it to destroy half of the Asgardian fleet, redirect the energy of Iron Man's missiles, and shoot pulses of concentrated energy.

THE SCIENCE OF MARVEL

In other entries, we have explored how energy production can be harnessed to create shocks (Hulk's Thunder Clap), mutate DNA (Gamma Radiation), fuel forges (Forging Stormbreaker), and collapse stars (Wormholes and Teleportation). If we were to measure the largest potential output of power, we would have to consider nuclear reactions to be among the most powerful reactions that yield energy. This energy is governed by Einstein's famous $E = mc^2$ equation, where E = energy, m = mass, and c = speed of light. But how exactly does this equation plug into the energy production happening at the center of the Power Stone?

Every element has an atomic number (equivalent to the number of protons) and an atomic mass (equivalent to the number of protons + neutrons). By typical convention, we often consider that the mass of a proton = mass of a neutron and a given atomic mass is simply two times the atomic number. However, if you paid extra close attention to the decimal points of the atomic number, you would realize that this doesn't add up. That small difference is called the mass defect and represents the mass (m) that is converted into energy that holds neutrons and protons inside the nucleus (E). We calculate this binding energy using $E = mc^2$. When an unstable element changes the composition of its nucleus or nucleons (protons + neutrons), it has to give off that mass defect in the form of energy. This offers incredibly powerful energies within the nuclei of atoms, which Thanos harnesses.

If Thanos's use of the Power Stone allows him to manipulate the nucleons of an element he can generate fission and fusion reactions. Fusion relies on the energies emitted when atoms fuse (as seen with hydrogen fusion in the core of the sun), whereas fission relies on the energy emitted when larger atoms split into smaller ones (as seen in uranium-235 decay in nuclear reactors). If Thanos used the Power Stone to change the number of nucleons in an atom he could use any element around him to carry out a nuclear reaction. Elements lighter than iron would fuse, creating energy, and elements heavier than iron would split, creating energy. Iron acts as a boundary between the duality of fission and fusion since the energy required to fuse iron can only be generated in supernovae.

WHAT ABOUT THAT PURPLE GLOW?

While it is difficult to imagine what atomic fuel is being fused/ fissioned when Thanos uses the Power Stone, we know that different elements can burn light of different colors. For example, potassium burns a violet/purple color that can be measured by flame emission spectroscopy. This burning color is the by-product of chemical radicals that excite the atom in question by increasing the energy of its electron to an unstable state that requires the release of energy. This energy is dissipated as light of different wavelengths that depend on the energy level(s) of those specific electrons. Or maybe Thanos and Ronan chose to use the Power Stone to only carry out fission reactions where the resulting decay products had potassium that could burn purple.

THE SCIENCE OF REAL LIFE

Perhaps the most historically relevant example of nuclear energy was seen during World War II when atom bombs were used by Allied forces against Japan. These two bombs hit Nagasaki and Hiroshima, killing approximately 130,000 people (most of them civilians). These bombs, code named Fat Man and Little Boy, used plutonium and uranium to generate a nuclear chain reaction. Consider how the uranium decay in Little Boy was able to produce the energy that decimated Hiroshima. In Little Boy, uranium-235 got hit by a neutron provided by a polonium trigger, and turned into the unstable uranium-236. This isotope immediately decayed into krypton-92 and barium-141 and released three neutrons. Those three neutrons each started another uranium-236 decay, tripling the energy emission and decay each time. Along with neutron reflectors and other ways to optimize the nuclear chain reaction, a tremendous amount of energy was produced.

The next step up from an atom bomb is a thermonuclear bomb, which thankfully has never been used in an attack between countries. This bomb relies on fission to achieve high enough temperatures to allow fusion of hydrogen. This fusion reaction is then used as a positive feedback reaction to carry out more fission. This bomb's design uses up the fission fuel more effectively than an atom bomb where the fissile products are not entirely decayed.

We have also learned to use uranium fission to fuel nuclear reactors and we hope one day to be able to control fusion as well (see Iron Man's Power Reactor). Nuclear fission is currently the only usable form of nuclear energy that became a viable interest after the bombs dropped in the 1940s. However, it wasn't until oil prices skyrocketed in 1973 due to war in the Middle East that nuclear energy received the necessary commercial and federal investment to start building reactors. The most common reactor used was the light-water reactor, which is made of a moderator, fuel rods, and control rods. Fuel rods are composed of uranium-235 ready for decay, and both moderator and control rods are used to control the fission reaction. A moderator is often just water or heavy water, which slows down neutrons emitted from fuel rods. Similarly, control rods made of boron or cadmium soak up free neutrons. Throughout the fission reaction water is heated and pumped through a heat exchanger that generates steam to spin turbines that generate electricity.

What about that purple glow? Flame emission spectroscopy is an important analytical chemistry tool that can allow us to identify trace elements in different samples. Dr. Ryan Anderson at the US Geological Survey's Astrogeology Science Center uses the Chem-Cam laser on the Mars *Curiosity* rover to zap rocks and use their emission spectra to determine the composition of the terrain on Mars. This laser can fire at a 1-mm^2 patch of land up to 7 meters away and uses a supersensitive onboard camera to determine many aspects of a soil or rock sample. For example, emission spectra can allow Anderson to determine whether rock is sedimentary or volcanic, to measure the abundance of all chemical elements, and to recognize ice and minerals with water molecules in their crystal.

Chapter 9

Fantastic Physics

WORMHOLES AND TELEPORTATION

WHEN: *The Avengers, Captain America: The First Avenger, Thor, Thor: The Dark World, Thor: Ragnarok, Avengers: Infinity War, Avengers: Age of Ultron*

WHO: Loki, Thanos, Red Skull, Heimdall

SCIENCE CONCEPTS: General relativity, subatomic particles

INTRODUCTION

Technology has provided us with many ways to speed up travel. We have commuter systems to cover small distances and planes to cross over countries. However, as we escape Earth's atmosphere and travel to other planets, timing becomes more of an issue. Imagine a flight to Mars which can take anywhere from one hundred to three hundred days depending on the alignment of Mars with Earth and the speed of your spacecraft. If you were to travel to our nearest neighboring galaxy, the Canis Major dwarf galaxy, it would take only twenty-five thousand years (if you could travel at the speed of light, which you probably can't). So what if there was another way to cover these distances in an instant? Perhaps it would involve possessing the Space Stone or an Asgardian uru weapon that can summon the Bifrost Bridge!

BACKSTORY

Throughout the MCU, the ability to travel between two points in space is facilitated by wormholes, Asgardian relics, and the Space Stone. Asgardian weapons forged from uru such as Hofund or Stormbreaker harness the ability to create rainbow portals, which can transport anyone across the universe. In the invasion of New York City in *The Avengers*, this is carried out by Loki, who uses the Tesseract to open a portal between incoming Chitauri alien forces and Earth. When Thanos destroys the Tesseract to reveal the Space Stone, he sets it in the Infinity Gauntlet and is granted the ability to instantly open portals to Earth, Vormir, and Titan. Lastly, wormholes like those created during the Convergence in *Thor: The Dark World* or the Devil's Anus in *Thor: Ragnarok* are used as a means to travel across various parts of the universe.

THE SCIENCE OF MARVEL

The creation of wormholes in the MCU is a source of incredible power. Heimdall, who is tasked with overseeing all Nine Realms, spends several millennia guarding this power and using it in service of Asgard. Similarly, in the hands of Loki and Thanos, the Space Stone wills into existence the creation of universe-spanning portals that eliminate the need for spaceships. The science considered for this type of travel was pitched as an Einstein-Rosen bridge by S.H.I.E.L.D. agent Phil Coulson and Bruce Banner whenever they were observed. So what exactly is happening inside the Devil's Anus, the Bifrost Bridge, and the many portals opened by the Space Stone?

EINSTEIN AND WORMHOLES

When Albert Einstein came up with his theory of general relativity, he developed a theoretical framework that could, in theory, allow for the formation of wormholes in space. This idea came to be during Nathan Rosen's work with Einstein between 1934–1936 as solutions to the Einstein field equations in 1938.

We can speculate that these relics could create an Einstein-Rosen bridge by understanding the formation of black holes. A black hole is a remnant of a large star's collapse (see Forging Stormbreaker). It's a point in space-time that generates so much density that gravity sucks everything into it, even light. Due to the mass and density of this black hole and Einstein's theory of general relativity, it also has the ability to affect the surrounding space-time, creating an infinitely dense singularity or, in the case of a wormhole, a path to another point in time-space. This exit point of the black hole is called a white hole, where matter travels outward similar to the expansion known as the Big Bang. However, these wormholes are incredibly unstable and gravity would cause them to instantly collapse. So the power of the Space Stone and Asgardian uru weapons could facilitate the opening of a wormhole but would also need to be able to exert another force to keep them open. One possibility would be to imagine that a stream of exotic matter, with negative energy, is being funneled into a black hole to keep it open. This exotic matter would repel the forces of gravity trying to close the wormhole and sustain the opening long enough for it to be traversed. We can speculate that this exotic matter was produced by the Tesseract and its blue stream of energy in *The Avengers* on the top of Stark Tower. Since most of the black holes we have learned about require stellar collapse, their formation would be a cataclysmic event that would do more harm than good. One alternative mechanism would function analogous to gaining access to micro black holes, which rely heavily on quantum mechanical effects. These black holes were formed during the early stages of the universe after the Big Bang where pockets of high density could cause the localized collapse of gravity. Since these black holes did not form from the death of a star, they can be as small as a human cell or large enough for Thanos and friends to walk through.

THE SCIENCE OF REAL LIFE

The idea of traveling through an Einstein-Rosen bridge is pretty impractical given our current toolset and knowledge. For example, could we find or create a black hole of the appropriate size? Could we keep it open? Could we successfully pass through it? Unfortunately, we are limited to our theoretical interpretation of this phenomenon. However, this hasn't stopped us from trying! A good start is to focus on the part of the black hole that would immediately kill us (it's usually good to start here). If you were to approach a typical black hole, it would generate an infinitely powerful gravitational force that would result in spaghettification of your entire body into the diameter of a single one of your atoms (see Forging Stormbreaker). This is a pretty wicked way to go, but what if we could reinterpret this?

In a study published by Dr. Diego Rubiera-Garcia at the University of Lisbon, the idea of a naked singularity was proposed to provide safe passage for an observer passing through a black hole. We assume that death waits for us inside a black hole because the event horizon encloses the singularity of gravity that cannot even let light escape. It is the space-time point of no return (even if you are traveling at the speed of light). However, Rubiera-Garcia and colleagues suggested that if our universe is made of five or more dimensions it may support a naked singularity. This would allow for a gravitational singularity with no event horizon and a wormhole big enough to allow an observer to enter it without falling victim to spaghettification.

The team even carried out simulations in which objects held together by physical or chemical interactions would pass through the wormhole with these interactions intact. In essence, the gravitational singularity would compress matter to the limits of the size of the wormhole itself and not into the infinite abyss.

Outside of the Einstein-Rosen bridge, we've also tried our best to model aspects of wormholes on Earth. In a study published by Dr. Alvaro Sanchez's team at Universitat Autònoma de Barcelona, a magnetostatic wormhole was created in the lab. Don't get too excited! It does not allow matter to pass through it, but it does emulate what a magnetic field looks like entering a wormhole. This setup takes a magnet and uses magnetic metamaterials to make it appear as if two magnetic monopoles are existing in empty space. This magnetostatic wormhole makes the center of a bar magnet invisible to magnetic sensors. The metamaterial cloak is clearly visible as a metallic sphere, but if you had a pair of glasses that could only detect magnetic fields you would effectively see two ends of a magnetic field suspended in space with nothing between them. It won't let us travel over time or space, but it can provide a model to understand how certain properties of real wormholes function, or at the very least how we can shield stuff from magnetic fields. For example, MRIs use a strong magnetic field to visualize your internal anatomy but make it difficult to image in the presence of metals. These magnetic metamaterials could help shield these effects.

PHASING

WHEN: Ant-Man, Captain America: Civil War, Avengers: Age of Ultron, Avengers: Infinity War, Ant-Man and the Wasp

WHO: Ghost, Vision

SCIENCE CONCEPTS: Quantum mechanics

INTRODUCTION
Matter obeys different rules at different scales. For example, a kilometer-wide cube of water floating above your head will fall to the ground much differently than a 1-millimeter cube of water. Scale also changes the way matter behaves at the subatomic level. For example, if you could throw a ball at a wall at your current scale you would expect the ball to bounce and gravity would make it hit the ground. However, in subatomic space, there's a chance that the ball may end up on the other side of that wall. How can we begin to explain this type of phenomenon, and can we use it to understand how Ghost and Vision can phase through walls?

BACKSTORY
In the Marvel Cinematic Universe, the ability to pass through walls is possessed by the android Vision and Ghost, a S.H.I.E.L.D.

operative. This skill involves changing density, in Vision's case, or quantum tunneling, for Ghost. In *Captain America: Civil War*, Vision used his phasing ability to pass right through a giant-sized Ant-Man, and in *Avengers: Age of Ultron*, he used phasing to destroy Ultron's legion from the inside out. In *Ant-Man and the Wasp*, Ghost was a trained assassin with the ability to sneak into and out of any situation using her phasing. Her ability to control her body's quantum waveform allowed her to incorporate phasing into a fighting style that never let her get hit, even when punched. Both characters also probably made very creepy roommates, not knowing how to knock or use doors.

THE SCIENCE OF MARVEL

To come close to understanding the phasing that Vision and Ava Starr (Ghost) can carry out, we need to consider how matter behaves in subatomic spaces. This goes into the realm of quantum mechanics and the ability of matter to behave as both a particle and a wave. While it is hard to wrap your head around this, if you consider quantum models for matter it is technically possible for a particle to be both. However, the flexibility of that model changes as you get bigger. Imagine if you will, the edge of Hope van Dyne's heel turning into a roundhouse kick swiping at Starr's head in their first encounter in *Ant-Man and the Wasp*. Typically, the force of her kick would be equivalent to the momentum (force = mass × velocity) of the hardest part of her heel. Instead, Wasp's leg blurs through Ghost's head and then Ghost completely disappears.

This phenomenon is similar to quantum tunneling, which occurs at atomic and subatomic scales. However, this doesn't entirely rule out the possibility of it happening at a larger scale. Quantum mechanics posits that matter exists in a blurry continuum of probabilistic states that only become determined once one of these states becomes observed. An object's probability to exist in a given state over time is coded in its wave function, and this wave function is determined by its De Broglie wavelength (λ). This can be calculated by dividing the Planck constant by the momentum of an object. The smaller the wavelength, the more defined the position and the less the uncertainty (Wasp's kick), whereas a larger wavelength is a less defined position and is more uncertain (Ghost's head). If we had to speculate, it would make more sense if Ghost's phasing ability relied on her ability to control her own De Broglie wavelength by manipulating the variables that define her own physics.

DE BROGLIE WAVE FUNCTION

Just to show you the math: in a fight between two normal people fighting in cosplay their De Broglie wave function would be defined as follows.

λ = Planck Constant/Momentum

λ_{Ghost} = (6.62607004 × 10^{-34} Joule·seconds)/(~56 kg$_{mass}$ × ~0.05 m/s$_{dodge\ velocity}$)

λ_{Wasp} = (6.62607004 × 10^{-34} Joule·seconds)/(~56 kg$_{mass}$ × ~0.78 m/s$_{kick\ velocity}$)

The Planck constant is a value that represents the connection between the wavelength of a particle and its quantization and is super tiny. Either way you look at it we have a very small wave function that defines a near absolute likelihood that both Wasp and Ghost would exist in the same space during the impact of the kick (with Ghost having a slightly higher, but ultimately negligible, amount of uncertainty). We know that this is not the case and the only way Ghost can increase her De Broglie wavelength is by dramatically decreasing her mass to that of a subatomic particle and/or slowing down her resting movement to a near standstill. This change in her De Broglie wave function is partly illustrated by the many shadows of herself that are seen throughout *Ant-Man and the Wasp* and during battle. Once she achieves this level of uncertainty her ability to pass through a barrier such as a roundhouse kick becomes significantly likelier. When she

CHAPTER 9: FANTASTIC PHYSICS

uses this power to pass through a barrier like she repeatedly does during combat, she is carrying out quantum tunneling.

THE SCIENCE OF REAL LIFE

If Ghost couldn't decrease her mass to that of a subatomic particle or slow her movement to a complete stop would there still be a chance she could phase through the Wasp's kick? Yes. Let's calculate the likelihood of this phase using the Schrödinger equation. The fundamental equation he came up with in 1925 earned him the 1933 Nobel Prize in Physics; however, it involves some calculus that is beyond the scope of this book. If you are one for partial derivatives and a deeper understanding of quantum mechanics, use that curiosity to seek out a graduate degree in physics!

SCHRÖDINGER'S EQUATION

In a simplified form, we can present the solution to Schrödinger's equation for the probability of quantum tunneling as $P = e^{(-KL)}$, where P is the probability, K is the wave number, and L is the width of the barrier. Before we can solve this, we should probably unfold the wave number as $K = \sqrt{\dfrac{2m(V-E)}{h^2}}$.

Here m = mass, V = potential of the barrier, E = energy of the moving object, and h = Planck's constant.

$K = [sqrt(2(4.5kg_{head}(\{4.5kg_{head} \times 9.8m/s^2_{gravity} \times 1.67m_{height}\}-\{^{1}/_{2}\ 1.36kg_{foot} \times 0.78m/s\ _{speed}^2\}))]/(6.62607004 \times 10^{-34}\ J \cdot s)\ _{Planck\ Constant}^2$

$K = 2.869402 \times 10^{44}$

Plug that into $e^{(-KL)}$ where our wave number is considerably large due to the large mass and energy differential between standing and moving objects. If we were to assume 0.05 meters for the thickness of the Wasp's foot:

$P = e^{(-(2.869402 \times 10^{44})(0.05))}$

$P = e^{-(2.869402 \times 10^{43})}$

$P = $ (negligibly just above zero)

Basically the likelihood of phasing through the Wasp's kick has an infinitesimally small chance of occurring. While still not impossible, it's more likely that you'd win the lottery every day of your life until the day you die and subsequent generations of your kin would do the same than that you'd phase through a kick. Frankly, I'd take winning on just one day over *maybe* not getting kicked in the head. What's helpful about this thought experiment is that for us to consider the quantum phenomena that occur in our universe we need to think smaller. When you do, you'll realize that quantum tunneling is pretty common and is necessary for our sun to shine the way it does.

At the center of the sun's core, hydrogen is fusing to make helium, which is fusing to make heavier atoms, etc. (see Power Stones and Nuclear Fission). However, the amount of energy we have to invest into starting fusion on Earth is significantly higher than at the core of the sun. When two hydrogen atoms get closer, they repel each other. In order to overcome this repulsion, the sun would need to produce a much larger energy input than is measured at its core. A shortcut around this energy investment is allowing these atoms to quantum tunnel through each other to fuse into deuterium. Despite quantum tunneling's rare occurrence, when you consider how much hydrogen is involved and how long the sun has been burning (4.603 billion years), it becomes more evident how quantum tunneling allows our sun to shine. This phenomenon exists throughout the universe, from the sun's atomic interactions to how the properties of silicon transistors store the 1s and 0s on a circuit board.

SHRINKING INTO THE QUANTUM REALM

WHEN: Ant-Man, Ant-Man and the Wasp
WHO: Ant-Man, Wasp
SCIENCE CONCEPTS: Quantum mechanics, fractal cosmology

INTRODUCTION

Our frame of reference in our universe is limited. This isn't our fault; we are simply the product of our evolution. This path has allowed our eyes to see only certain wavelengths of electromagnetic radiation, exist at a certain scale, and age to a certain time. We've gotten clever enough to widen our thinking to abstract ourselves relative to the scale of the planets, stars, and galaxies that exist in our universe. As we scale down, we've also been able to abstract what exists beyond what we can see. We begin to understand how life exists at every scale, from the blue whale all the way to microscopic microorganisms. As we delve even deeper we go into the "spooky" physics (as Einstein put it) of quantum mechanics. Which raises the question: what would this look like to a hero like Ant-Man?

BACKSTORY

In *Ant-Man* and *Ant-Man and the Wasp*, the secret of Pym Particles is the ability to unlock access to the Quantum Realm. However, before getting there, they take a fantastic journey through scale. Some of these changes in scale don't necessarily change the way in which the world is perceived (e.g., toddler-sized Scott Lang in *Ant-Man and the Wasp*). On the other hand, as Lang and Pym take their journeys to become smaller, they begin to see insects, the rough texture of smooth objects, planet-sized dust particles, single-celled life, atoms, and subatomic space. Once they enter this subatomic space they take their first foray into the Quantum Realm, seeing symmetrical reflections of themselves before falling into a dark void with dimly lit spaces.

THE SCIENCE OF MARVEL

In the MCU, the journey into the Quantum Realm is fueled by Pym Particles circulating in the self-contained closed Ant-Man suit. The first time we see this is when Lang needs to shrink "between the molecules" and through the titanium metallic bonds of the Yellowjacket armor. In the same journey in *Ant-Man and the Wasp*, Henry Pym shrinks using his Quantum Vehicle to sizes where he witnesses some of the smallest animals trying to digest his vehicle. These animals, called tardigrades, measure about 10^{-4} m, suggesting he had already shrunk smaller than the average size of a cell ($\sim 10^{-5}$ m). Eventually, both Lang and Pym begin to observe bacteria measuring at 10^{-7} m. As Lang and Pym get to about the size of a virus, they start to shrink smaller than the wavelengths of the visible spectrum of light. If they shrink below 600–700 nm they stop seeing any orange and red. As they get even smaller, yellow, green, and blue fade away, and when they shrink smaller than 400 nm, everything goes dark. In this dark space, they are the size of a virus (10^{-9} m), and biomacromolecules like DNA and protein start to reveal themselves including their atoms and bonds.

As both Pym and Lang approach the size of an atom (10^{-10} m) they collapse into moving three-dimensional fractal geometry, resulting in their entry into the Quantum Realm. (A fractal is a recursive never-ending mathematical pattern.) This is particularly interesting because it seems as if the MCU is leading its audience into the theory of fractal cosmology with moving fractal geometries. Fractal cosmology theorizes that our universe may exist within the subatomic space of the smallest particles of another universe, which exists within the subatomic space of another universe, and so on.

Once inside the Quantum Realm, Lang and Pym lose their minds—quite literally. At this scale, a particle no longer follows the typical rules of physics and begins to adopt a probabilistic model of "being." Until observed at a macroscopic level, Pym and Lang exist as a wave function, defined as the probability of existing at every single point of space at every single point of time. This is called quantum superposition and seems to be partly illustrated by Pym's quantum "shadows" drifting away from his body as he stumbles out of his Quantum Vehicle in search of Janet van Dyne. As a result, spending too much time in the Quantum Realm or the subsequent use of Pym Particles causes one to lose oneself in a probabilistic loss of identity.

When Lang falls into the Quantum Realm he also interacts with Janet van Dyne, causing their minds to wire across time and space through quantum entanglement. This suggests that through interactions in subatomic space, certain particles can be linked together through their quantum state. This typically involves a photon or electron with different atomic spins but, through the use of Pym Particles, suggests it has transitive properties to organs in a whole human (i.e., Lang's) brain. When van Dyne and Lang meet in the Quantum Realm as their subatomic selves, their interaction allows van Dyne to entangle her thoughts into Lang's neural processes. With the use of Pym Particles, this interaction remains in the brain of Lang long enough for van Dyne to use Lang as a channel to talk to her husband and daughter.

THE SCIENCE OF REAL LIFE

In real life, we don't have Pym Particles to defy the laws of reality. Sadly, we are restricted to the square-cube law, where the ratio between two volumes is always greater than the ratio between their surface area. This means as an object shrinks, its volume is decreasing faster than its surface area, leading to significantly higher density (if Lang is retaining his original weight). Long before Lang would reach the Quantum Realm he would have to be fusing the atoms in his body, causing a nuclear reaction (see Power Stones and Nuclear Fission). These physical laws make it possible for certain animals to carry multiple times their weight (see Giant Ants). But what about the spooky effects of fractal universes and entering the Quantum Realm?

The idea of fractal cosmology was first put forward by Dr. Andrei Linde to suggest that the universe is eternally inflating at various scales. However, a recent survey of two hundred thousand galaxies in a cubic volume of three billion light years has said otherwise. Instead of looking into the smaller scales of space like Pym and Lang, Dr. Morag Scrimgeour and her colleagues at the International Centre for Radio Astronomy Research in Australia looked outward. In their data set, called the WiggleZ Dark Energy Survey, they found that matter is distributed evenly across the universe. If we were living in a fractally derived universe, matter would be consistently clustered as we changed scale within the universe. This is kind of good news since if fractal cosmology was real, we'd have to rewrite Einstein's theory of general relativity and revise our understanding of dark matter and energy!

And what about quantum superposition? This idea has been illustrated by a famous thought experiment proposed by Erwin Schrödinger regarding his poor fictional cat.

SCHRÖDINGER'S CAT

In 1935, Austrian physicist Erwin Schrödinger asked the following question: if you put a cat in a box with poison that could be activated randomly by radioactive decay, is the cat alive or dead? The only way to find out is to open the box and see for yourself. However, the existence of the cat before you open the box is a probability defined by maybe dead/maybe alive. This is referred to as quantum superposition.

Quantum superposition describes the behavior of quantum objects defined by a wave function of its probability states. Once this quantum object is observed, it collapses the wave function to the observed outcome. This property is fundamental to the ability of particles to become entangled in different states. That is, knowing

CHAPTER 9: FANTASTIC PHYSICS

the state of one particle immediately gives you information on the state of another particle. We typically measure this on photons with their relative spin states, and we've been able to replicate this finding across incredibly large distances. In a study led by Dr. Juan Yin of the University of Science and Technology of China, his team was able to demonstrate the entanglement of photons separated by 1,203 km. This experiment used the Micius satellite and may lay the foundations of next-level encrypted communications. However, our current understanding of entanglement limits our ability to use quantum states to infer information that can send an actual message. Technically only upon observing the quantum state of a particle on both sides of entanglement can we verify a sequence of information. As a result, using this as a form of communication would still require a quantum computing language that doesn't require "checking our homework," since that would defeat the purpose.

WEB WINGS

WHEN: *Spider-Man: Homecoming*	
WHO: Spider-Man	
SCIENCE CONCEPTS: Aerodynamics of flight	

INTRODUCTION

In the animal kingdom several different species have evolved flight. Birds and bats modified their forearms to develop wings whereas some insects created outgrowths from their thoracic exoskeleton to form wings. Humans have even borrowed designs from the form and function of an animal's wing to engineer airplanes. However, flight did not always look so graceful. In its earliest evolution it involved fewer wingbeats and more gliding. How did these flight-related structures evolve, and how exactly do they manipulate airflow to generate lift and allow many flying creatures to travel long distances? Can these lessons be applied to help humans fly? A spider? Or even a Spider-Man?

BACKSTORY

When the Spider-Man franchise was rebooted for a third time and entered the Marvel Cinematic Universe a lot of things changed about the character. He became younger and more agile in combat with a costume throwback to the very first Marvel comic book issue in which he appeared, *Amazing Fantasy* #15. His costume included dynamic eyes that could squint and web wings trailing from his elbow to his waist. In *Spider-Man: Homecoming* his web wings were showcased in a death-defying leap to save his classmates in a falling elevator. In this scene, Spider-Man climbed to the peak of the Washington Monument, proceeded to backflip, twist, and spread his wings to clear a police helicopter, and used that momentum to break through a 4-inch ballistic glass window on the monument.

THE SCIENCE OF MARVEL

Spider-Man's main form of mobility is swinging like a pendulum on his webbing, and he spends most of his time using his falling momentum to swing all over New York City. However, the addition of web wings gives Spider-Man a deployable wingsuit when he doesn't have the luxury of grappling points. This was demonstrated in the Washington Monument elevator rescue where he was left with few places to swing to and he had to glide several meters in order to clear the distance required to swing with the necessary momentum to break the ballistic glass. Also, during typical travel Spider-Man's web swinging uses up his web fluid and slows down on the upswing of his pendulum-like motion. While this may be an effective way to allow Spider-Man to surveil New York City for criminals, it wouldn't be helpful if he needed to cover large distances as quickly as possible. In such a case, his web wings would give him a controlled fall and glide that would cover more distance and reach higher velocities (and definitely deliver a more powerful blow to his enemies!).

How exactly would this pseudo-fly suit allow him to glide over New York City? As airflow travels toward Spider-Man, his angle of attack determines the angle at which his web wings receive the relative wind. As Spidey lowers his angle of attack, he reduces his drag, allowing him to reach faster speeds. At a higher angle of attack he will have more drag to slow him down. His angle of attack and flight path are shaped by variables such as his chord line, pitch, and flight path angle. His chord line is defined as the plane at which both of his web wings are aligned to receive the relative wind, and his pitch is determined by his body's angle relative to the horizon. So when Spider-Man is jackknifing feet first during a fall his pitch is +90 degrees, and his angle of attack is 0 degrees. If he is free-falling with his body horizontal to the horizon, his pitch is 0 degrees and his angle of attack +90 degrees. Lastly, his flight path angle is the angle from the horizon to his flight path, or the vector that determines the direction he wants to travel. Spider-Man's angle of attack changes with relative airflow beating against his chord line with small changes to his pitch. Since most wingsuits have a glide ratio of 2.5, which is determined as height/distance, if Spider-Man needs to swoop in from the top of the Empire State Building (381 m) to tackle the Vulture a block away (154 m), he would take a leap that would decrease his angle of attack with a flight path angle of −45 degrees. During this fall he would pitch his body forward −30 degrees from the horizon, giving him an angle of attack of +25 degrees, causing incoming airflow to hit his web wings and generate lift. The lift he generates as he falls will create a force that will carry him the necessary 154 m.

THE SCIENCE OF REAL LIFE

Unlike the multimillion-dollar investment that Stark Industries made to ensure the safety of humankind by equipping the fictional Spider-Man, others use wingsuits that look like a flying squirrel Halloween costume for sport. Wingsuiting was a private endeavor for many until

1999 when the first commercial vendors (BirdMan International Ltd.) began producing their own brand of wingsuits and instituting their own training program. Those interested in this daring sport are required to have completed at least two hundred skydiving jumps before qualifying for wingsuit training. Despite these safety precautions, a 2012 study examining the severe injuries that result from BASE (building, antenna, span, and earth) jumping showed that risk for severe injury increased significantly with the number of jumps. Within this BASE jumping cohort, 72 percent had borne witness to a death or severe injury.

FLYING SQUIRRELS

While it's possible (though silly) to engineer outfits that allow humans to glide, evolution has also given us natural gliders. For example, the North American flying squirrel and the Australian sugar glider have both independently evolved flaps of skin allowing gliding in order to colonize treetops. Flying squirrels have been known to glide 90 m. Sugar gliders can travel approximately 50 m by gliding.

Since we are talking about a gliding Spider-Man, what about gliding spiders? While some may not be keen to hear it, Dr. Robert Dudley and his team at the Smithsonian Tropical Research Institute have discovered spiders in the genus *Selenops* taking flight. In research that simply involved dropping fifty-nine spiders from tree canopies in Panama and Peru, Dudley's team found 93 percent of these spiders would direct themselves in a controlled fall to a nearby tree where they would land safely. These species of spiders have evolved flattened bodies and a falling behavior where they extend their front limbs forward to aid in a controlled descent. This illustrates a strong selective pressure against uncontrolled falls for these arachnids in arboreal environments. So if you ever thought one of these spiders fell on

       CHAPTER 9: FANTASTIC PHYSICS

your head by accident, be reassured by the thought that it totally intended to land in your hair.

If you're not bothered by falling spiders gliding in from above, then perhaps you would also be interested in spiders that fly upward from below. In a separate study led by Dr. Erica Morley at the University of Bristol in England, her team discovered that atmospheric electrostatic forces allow certain spiders to "balloon" and fly long distances. This behavior involves a spider shooting silk from its abdomen into the air, where it gets lifted up. The initial assumption was that wind was carrying these spiders upward; however, ballooning behaviors were mainly seen on days with gentle winds. Morley's group decided to see if atmospheric electrostatic forces could be playing a role by isolating spiders in a chamber where the researchers could control airflow and the electrostatic currents. When her team induced an electrostatic force similar to what would happen in the field, spiders began to tiptoe on a platform, raised their abdomens, and began ballooning behavior. Once ballooning behavior began, the spider could be lifted or dropped by turning the electric field on and off. Her group further showed that an electric current would cause the sensory hairs on the spider's exoskeleton to stand up, suggesting they could sense these electric currents, which led to the observed behavior (remember that scene in *Avengers: Infinity War* in which Peter Parker sees Thanos's ship flying over New York City). While swinging by spider thread is a favorite mode of transportation for Spider-Man, spiders are expert BASE jumpers and aviators in their own right.

THE EYE OF AGAMOTTO

WHEN: Doctor Strange, Avengers: Infinity War, Thor: Ragnarok	
WHO: Dr. Strange, Thanos	
SCIENCE CONCEPTS: Closed timelike curves, general relativity, time	

INTRODUCTION

Humans measure time using seconds, minutes, hours, days, years, etc. Time is one of the most precious things a human has in a lifetime. Time can be perceived very differently by two people. A one-hour lecture can feel like a thirty-minute discussion of something that really interests the lecturer or it can feel like a grueling five hours to a student who isn't engaged with the topic. Outside of our own perception, time is in fact relative and not consistent across every point of the universe. If time is relative, how can we reliably measure it? How do we take into account factors such as gravity that can bend space-time? And will we ever be able to exploit this understanding to travel into the future and the past?

BACKSTORY

In the events of *Doctor Strange*, the Eye of Agamotto is described as one of the most powerful magical relics created by the first Sorcerer Supreme. This relic encases the Time Stone, which allows its users to change the flow of time using magical incantations. In the first example of its use, Strange rewinds the clock on an eaten apple. Interestingly, the power of the Eye is limited to a bubble of time-space where its properties exist independent of its relative surroundings. Use of this relic has also been considered to be unwise since its effect can generate multiple timelines, disrupting the natural order of our reality. As such, it appears that Strange uses the power of the relic in situations where tremendous threats face all of reality (against Dormammu) or when he can safely observe the future in a form of clairvoyance (when Strange was predicting the future outcomes of the Infinity War).

THE SCIENCE OF MARVEL

In the concluding events of *Doctor Strange*, Wong and Strange mutually agree that Strange's possession of the Eye of Agamotto should be restricted. Since his knowledge of its power is limited it may be better to err on the side of caution and avoid starting alternate timelines that would disrupt the natural order of the world.

The idea of alternate timelines and universes seems to be congruent with Strange's control over "reality" and access to pocket dimensions. Nonetheless, the power of the Eye of Agamotto seems to be limited to the immediate physical space in which it is being used. For example, in Strange's bargain with Dormammu, he introduces time as a faint glow around his conversation with Dormammu, allowing him to reset time within this finite space. But why would this matter?

There are a few ways in which the Eye of Agamotto may work but at the very least it seems to be consistent with a causal timeline. That is, any change or interaction that happens due to time travel remains congruent with all outcomes that occur onward. The most logical way to confine a causal timeline is to keep it within space irrespective of time. This also appears to be how Strange confines the use of the Eye of Agamotto to a small bubble of time-space where he uses a form of time "observance" (clairvoyance) in the events of *Avengers: Infinity War*. (Let's even consider this multiverses observance.) In effect, the Time Stone may be allowing him to create a closed timelike curve (CTC) in his current fixed point in time and space (also, let's assume Marvel thought a fixed point in time-space was fixed even if it was on a rotating planet orbiting a star).

A CTC considers an object, such as Dr. Strange wearing the Eye of Agamotto, within four-dimensional time-space. Strange and the Eye follow a world line that tracks their space together through physical space (Strange positioned in space x,y,z in NYC Sanctum Santorum) and then traveling through that space in time (Strange moving from $X,Y,Z_{Sanctum\ Santorum} \rightarrow X,Y,Z_{Q\text{-}ship} \rightarrow X,Y,Z_{Titan}$). His world line defines his movements across time-space going in one direction (with time moving forward). However, if the Eye of Agamotto were able to generate an incredibly large gravitational field, it could bend the curvature of time-space within the local vicinity of its green aura. In Einstein's predictions about general relativity, this strong gravitational force can cause frame dragging. Frame dragging allows Strange and the Eye of Agamotto to maintain a non-static stationary distribution of mass-energy (possibly illustrated by Strange going into "convulsions" during his clairvoy-ance?). Within the bubble of time exerted by the Time Stone, grav-itational forces may be bending time-space for Strange to create a CTC where his own world line would go into the future, curve around, and return to the same place 16,000,605 times. This, in a way, would allow him to experience the various scenarios expected to happen before a suitable one accommodates the preferred out-comes. Strange basically became a time-space tourist.

THE SCIENCE OF REAL LIFE

In real life, the idea of time travel introduces several paradoxes, such as the grandfather paradox. This thought experiment puts forth the idea that if you were to go back in time to murder your grandfather, you would be preventing your own birth and thus the ability to kill your grandfather. This hasn't entirely ruled out the possibility of time travel but has made it a particularly difficult phenomenon to under-stand. One thing that is somewhat possible is to slow down time or speed it up, relatively speaking (literally). When Einstein came up with his theories of general and special relativity he created a framework of physical rules that the universe has to follow. Within this ruleset is the phenomenon known as time dilation, which can actually allow an object to travel into the future, relatively speaking.

Time dilation occurs when an object accelerates to the universal speed limit, the speed of light (c). In general relativity, nothing can travel faster than light. Throwing a ball on a moving train lets the ball have the combined velocity of the train and the ball. However, if you turn on a flashlight on a moving train, the speed at which the light is traveling remains the same—no matter how much the train speeds up. If everything is limited to the speed limit of light, an object accel-erating to the speed of light begins to experience time differently than a stationary object through the process of time dilation.

In one of the earliest experimental proofs of this effect, Richard Keating and Joseph Hafele set a clock in a plane that flew around the world twice. They then compared it to clocks on Earth. Disparities in the time were found among clocks that were accelerated to a higher velocity in the air and also depending on the eastward/westward direction, consistent with Einstein's theories. Further evidence can be seen in various satellites orbiting the Earth and the astronauts in the International Space Station. Relative to us, these objects in space are traveling much faster than we are and thus are moving at speeds closer to the speed of light (although only fractionally so).

Is there any evidence that a CTC can sneak past something like the grandfather paradox? In the very small space of subatomic particles, kinda. Like everything in quantum mechanics, new rules often need to get made when the world gets smaller (see Phasing). This is partly due to the way in which a small particle is sometimes defined by a probabilistic wave of possibilities that become infinitesimally small as that particle gets larger. This idea was formalized in a hypothetical solution to Einstein's field equations proposed by David Deutsch in 1991 suggesting that at the quantum level, a particle follows a path of self-consistency. This is analogous to the idea of a causal time loop where time travel facilitates events to transpire exactly as they did/should.

POLARIZATION

Deutsch's theory was tested in work carried out by Dr. Tim Ralph of the University of Queensland in Australia. In his simulation, he took a pair of polarized photons traveling through a CTC to flip a switch on the machine that produced it. Polarization was an effective means of encoding the identity of the quantum state of a photon to reidentify the traveling photon that entered and emerged the CTC. The researchers found that the photon that emerged was consistently the same one that went into the machine originally. These were all simulations, but Ralph effectively demonstrated that their mathematical simulation was robust enough to report their findings.

Chapter 10

Magnificent Marvels

IRON MAN'S POWER REACTOR

WHEN: *Iron Man, Iron Man 2, The Avengers, Iron Man 3, Avengers: Age of Ultron, Avengers: Infinity War*

WHO: Iron Man

SCIENCE CONCEPTS: Fusion

INTRODUCTION

When you plug an appliance into the wall you connect it to a network of electricity that has been implemented into the infrastructure of your town or city through transformers, power cables, and power plants. We typically use this energy in the form of electric current with an electrical potential; the energy can come from solar, thermonuclear, hydroelectric, and even geothermal power sources. Knowing what we know about how this energy is generated and stored, is it possible to create a battery that can supply 8 gigajoules of energy while fitting comfortably in the palm of your hand? Most importantly, how could we make this power source safe enough to implant several inches into your chest to power an iron suit?

BACKSTORY

In *Iron Man* Tony Stark is given a heap of scrap metal from old missile casings and their electronics parts and tasked by his captors to engineer Jericho missiles. Instead, he strips electronics for palladium that he melted down to make into rings that would be wrapped with wire, making a miniaturized arc reactor that would fuel his first Iron Man suit. In subsequent iterations, Stark uses the designs of his father's experiments with the Tesseract to create a new element that can generate even more energy in *Iron Man 2*. However, Stark isn't the only one who had access to these forms of unbridled power.

THE SCIENCE OF MARVEL

Stark is secretive about how to use this technology, fearing it might be used as a weapon or to power other iron suits that may fall into the wrong hands. However, there are a few clues that can allow us to deduce aspects of its design. For example, dialogue between Stark and Dr. Ho Yinsen explicitly states that the arc reactor functions primarily as an electromagnet to prevent pieces of shrapnel from further entering his heart. This aspect of the electromagnet explains the toroid (the surface created by a curved plane) enclosure of the reactor, which receives a current of electricity from its center to coil multiple times around the toroid. This number of coils when a current passes through it creates a stronger magnetic field that attracts the pieces of shrapnel. This also means that the magnet probably needs to be on all the time and/or may act like an energy sink for excess energy from the core. This doesn't directly regulate the generation of electricity but works as a secondary function to remedy a symptom of his injury.

Stark can seemingly harness electrical energy directly from the arc reactor. Many other forms of energy usually need to be converted into electricity. For example, a dam uses the potential energy of water falling downward to spin a turbine that in turn spins magnets around copper coils generating electric energy. Since these other forms take up tremendous amounts of space and/or generate high levels of heat, we can propose that the arc reactor is facilitating nuclear beta decay. Elements in the center of the reactor core are functioning as a betavoltaic cell, liberating electrons into a form that can be directly applied to an electric load. This can be carried out using radioisotopes such as tritium (^{3}H) or, more conveniently, palladium (^{107}Pd). The decay of ^{107}Pd results in the loss of 1 neutron and subsequent generation of ^{107}Ag (an isotope of silver) and 1 electron that can sustain a supply of electricity. If you really wanted to, this beta decay can even excite different phosphors painted onto its containment vessel to glow blue, but this would be more about showing off.

In Stark's second iteration of the arc reactor he trades in his (possibly) betavoltaic cell and toroid enclosure for an all-new element, one discovered by his father, Howard Stark, during his research on the Tesseract. The element prompted Tony Stark to build a DIY particle accelerator in his home. This circular particle accelerator was used to create a beam of subatomic particles (electrons or protons) or whole atoms of elements that could be used to generate either new elements or isotopes. While we are uncertain what particle source Tony Stark used, it was fed into the vacuum of the circular accelerator where it was sped up with a series of electromagnets. Between these electromagnets, electric fields switching between positive and negative generated frequencies of radio waves that bunched up the particles along the circular accelerator. Whereas most scientists exercise caution in putting a target material in the path of the circular accelerator, Stark breaks the beam's path by steering it through his home, hitting a metallic triangle. This new mysterious element provides even more power for Stark's Iron Man armor and probably undergoes beta decay like the isotope ^{107}Pd so it remains compatible with his suit's technology only with fewer toxic effects.

THE SCIENCE OF REAL LIFE

Realistically it would be pretty difficult to create the base components of an arc reactor in the middle of a desert, even if you are in the scrapyard of a terrorist camp. ^{107}Pd is a trace isotope, meaning it would be very difficult to find, even in a weapons scrapyard. Even so, if you were to find it, there would be no way to tell it apart from any of its six other isotopes. It also wouldn't make sense to keep wearing an electromagnet if you could hire a good surgeon to remove the shrapnel when you returned to New York City and, you know, avoid the heavy metal poisoning Stark suffered in *Iron Man 2*.

While it's probably not possible to miniaturize a reactor that can supply so much energy to fit snuggly into your chest, it hasn't stopped engineers and physicists from coming up with newer and cleaner forms of energy. Current conventional forms of energy are not as effective as we want them to be. For example, thermonuclear energy requires large power plants that need to effectively manage nuclear fission and the dangerous amounts of heat released, and the resulting nuclear waste presents disposal problems. Solar radiation is a luxury afforded to parts of the world where it's usually sunny, and it doesn't produce high enough energy yields to power an entire city. To find new solutions to these issues, we have turned to new forms of cleaner nuclear energy.

One form of clean energy research showing great promise for the future is a system called magnetic confinement fusion that involves a doughnut of shining hot plasma. This process heats fast-moving hydrogen particles in a torus to six times the temperature of the sun's core. At this temperature, hydrogen atoms begin to fuse together and create hot plasma. As you might imagine, the process is pretty chaotic and can destroy the material container. However, applying strong magnetic fields to the plasma can confine it safely by forcing fusion products and atoms to spiral around magnetic field lines. Unlike thermonuclear reactors, it can be fueled by abundant resources on Earth and does not produce any radioactive or toxic by-products. The main hurdle to be overcome is that the amount of energy that needs to be invested into the confinement and generating of the plasma currently exceeds the amount of energy that can be generated by hydrogen fusion. Once this hurdle is overcome it could pave the way to a self-sustaining loop of plasma that only needs to be fed fuel to continue the reaction.

Currently, this technology is being used for several experimental reactors at MIT, Princeton, and the Max Planck Institute. There is an international effort by the European Union and six other countries to build an International Thermonuclear Experimental Reactor (ITER). This reactor is expected to be the world's largest thermonuclear reactor. It will hopefully achieve the burning of plasma and generate 500 megawatts over the course of four hundred seconds with only 50 megawatts of input energy. The ITER is expected to be completed in 2026.

SHAPESHIFTING

WHEN: Captain America: The Winter Soldier, Captain Marvel, Agents of S.H.I.E.L.D.

WHO: Black Widow, Skrulls

SCIENCE CONCEPTS: Cellular biology, animal pigmentation, materials engineering, genetics, biomimetics

INTRODUCTION

As either predator or prey, members of the animal kingdom have evolved many strategies to become cryptic or conspicuous. For example, variations in pigmentation patterns are plastic in a cuttlefish and allow it to become almost invisible in its natural environment. These changes occur neurally within seconds using cellular mechanisms that change both color and texture. Understanding how a cuttlefish can achieve this feat can offer insight into the biomimetic engineering of smart materials that can change how a human looks. How exactly does this occur in a living organism, and can we translate these biological phenomena into "shapeshifting" technology? Could these changes be so convincing that they would let HYDRA take over S.H.I.E.L.D.? Or even give an alien species the means to secretly invade Earth?

BACKSTORY

In the MCU, many covert operations are carried out by impersonating key personnel using a photostatic veil. In *Captain America: The Winter Soldier*, Black Widow uses one of these veils to assume the identity of Councilwoman Hawley and stop Alexander Pierce from taking hostages during Project Insight and his secret takeover of S.H.I.E.L.D. This veil, activated by a tap on her temple, changed several of Black Widow's facial features and even the sound of her voice. This technology is further used by HYDRA (Sunil Bakshi and Agent 33) in the *Agents of S.H.I.E.L.D.* TV series to carry out various cloak-and-dagger missions. Lastly, the ability to shape-shift is also introduced as a key physiological ability of the Skrulls in the events of *Captain Marvel*.

THE SCIENCE OF MARVEL

In order to break down the speculative science of shapeshifting, we will examine the case of a Skrull and the photostatic veil, separately. Both ideas have similar outcomes but use completely different substrates (i.e., a holographic cell vs. a patch of skin). Unlike humans, Skrulls—non-mammalian vertebrates and invertebrates—have various coloration patterns that they derive from their ability to produce diverse pigment-bearing cells called chromatophores. Whereas humans have melanocytes that deposit melanin into our skin to darken it, a cuttlefish has a variety of white, iridescent, orange, and red chromatophores existing in different layers within and below the skin. Each chromatophore exists between two states: aggregated or dispersed. At a microscopic level, aggregated chromatophores migrate all of their pigments to the very center of their cells, masking this color at a macroscopic level. In contrast, dispersed chromatophores migrate all of their pigments to the farthest extremities of the cells, causing this color to be more conspicuous at a macroscopic level. The overlapping of chromatophores in dispersed or aggregated states of the same color or different colors allows the generation of vibrant colors that can be observed by eye.

At a resting state a given Skrull appears green for the same reason a frog appears green. On its skin it has three layers of chromatophores that have varying degrees of dispersal and aggregation. On the basal layer, it has dark pigment–containing cells (melanophores), in the middle layer it has reflective cells (iridophores), and on the top layer it produces yellow pigment–bearing cells (xanthophores). As light passes through the skin it gets reflected blue by the middle layer of iridophores into the yellow xanthophores, making what appears to be green. It is possible Skrulls use this cellular repertoire of chromatophores like living pixels, changing their size to generate a palette of different human skin tones. However, this only makes a human-colored Skrull. What about the other features of a human's face? Like a cuttlefish, Skrulls may also use hydrostatic structures on the surface of their skin called papillae to deform and change their morphological texture to develop the more nuanced features of a human's face. These abilities require a degree of neurophysiological control; they are controlled by innervating the skin and neurotransmitters secreted from different glands in the Skrull's body.

In a similar fashion, the photostatic veil uses miniaturized holographic cells in a textile array that adheres to its user's face. Each holographic cell projects a holographic voxel that makes up the shape and color of a simulated face.

MIMICKING THE VOICE

Preprogramming a veil to assume a given identity requires a voice sample and DNA. Unlike the formation of textured papillae in Skrulls, the photostatic veil likely sequences and sources a given genome into the cloud where it computationally predicts facial structures by cross-referencing variations in DNA sequence with different facial characteristics in a much larger database of faces with sequenced genomes. Adjustment to voice can be made by typical voice-changing software that can change pitch amplitude and phase of sound waves and filter them through the nuanced features of that person's speech.

CHAPTER 10: MAGNIFICENT MARVELS

THE SCIENCE OF REAL LIFE

As the animal kingdom has shown us, cuttlefish have developed a mastery over camouflage since they use it to hunt, hide from predators, and communicate with other cuttlefish. However, these talents are not restricted to cephalopods. In my research with the African cichlid fish *Astatotilapia burtoni* (see Hulk's Transformation), I have observed changes in its color with the expression of genes that regulate the sensitivity of chromatophores to disperse/aggregate yellow pigments. Several species of frogs, lizards, and crabs make these changes using similar cellular mechanisms. Unfortunately, this plastic trait is unlikely to be one that humans will ever possess. One of the reasons we have not evolved diverse chromatophores may have to do with the nocturnal patterns of our mammalian ancestors 225 million years ago. At this stage in our evolution, cold-blooded dinosaurs in the Triassic Period required diurnal activity to keep warm, while our shrew-like ancestors survived because they were nocturnal and hidden. In a nocturnal niche, vision was not as important for our ancestors as it was for diurnal species. This lasted until about sixty-five million years ago as mammals began to fill in diurnal niches, but at this point reptiles and fish had a significant head start.

What about using technology to artificially simulate these powers? While holograms do physically exist (see Manipulating Reality), the technology to produce them has not been miniaturized to the degree that it can be stitched onto a textile. The closest analog of such a technology wouldn't use holograms but biomimetic materials that can be filled with pigments. This has been carried out by Dr. Jonathan Rossiter of the University of Bristol. In his work, he has been able to use dielectric elastomers filled with pigment to mimic the function of chromatophores. This elastomer acts like a synthetic muscle that contracts when an electric current runs through it, causing the pigments to shrink away. While the current prototype melanophores that his lab has made are larger than a dermal cell, the technology could be miniaturized into soft robotic dermal layers that could generate various pigmentation patterns. These materials may be particularly helpful for dynamic camouflage and thermoregulation; for a nano mask you may be better off using a good foundation.

One interesting aspect of the photostatic veils programming requires DNA to determine the shape and form of a face. While not as robust as the technology at S.H.I.E.L.D., this is possible to determine from a genome. This field of study is called DNA-phenotyping and requires large screens of human genomes and correlating variations across the genome to topographical maps of the human face. In a 2018 study led by Christoph Lippert at Human Longevity Inc., 1,061 sequenced genomes and detailed measures of physical traits were enough to develop a computational framework for identifying a human by his or her genes. This work used individuals from mixed ancestries and was able to reidentify individuals eight out of ten times. However, this algorithm was not as effective when examining individuals with similar ancestral backgrounds. While this work may not be used to develop face-stealing technologies, it does raise several ethical concerns related to the privacy of genetic data!

VIBRANIUM

WHEN: *Captain America: The First Avenger, Captain America: The Winter Soldier, Captain America: Civil War, Avengers: Age of Ultron, Black Panther*

WHO: Ultron, Captain America, Black Panther

SCIENCE CONCEPTS: Piezoelectricity, metallic bonds, materials engineering

INTRODUCTION

We've engineered many elements to take on different molecular shapes that imbue it with macroscopic abilities. For example, carbon can exist naturally in a crystal lattice to make diamonds, in layers to make graphite, or in a more disordered form to resemble coal. Despite each material being made from the same element, they all have different electrical, physical, and chemical properties. Artificially, we can even change the molecular structures of carbon into nanocarbon tubes with tensile strengths and electric conductivity that would never be found naturally. Within these molecular structures, what would it take to repurpose a metallic element into a form that can absorb vibrational energies or become an indestructible alloy?

BACKSTORY

The most exotic and rare metal known to the human race in the MCU is vibranium. The world's supply of this metal fell from the sky thousands of years ago and exists in finite amounts under Mount Bashenga in the nation-state of Wakanda. While explosive and volatile in its raw processed ores, vibranium can be made into the most powerful metal alloys, textiles and energy supplies. Vibranium-steel alloys have been used to make Steve Rogers's shield nearly impenetrable and perfectly balanced. Technological aspects of vibranium in Shuri's nanites allow the Panther habit to absorb kinetic energy and store it for deadly purple shock waves in battle. So, what can we infer about vibranium's chemistry and physics that would allow us to understand how it can be so versatile in function?

THE SCIENCE OF MARVEL

To outsiders, Wakandan technological advances are often attributed to their access to vibranium ores, but there is clearly much more to it than that. It is impossible to imagine a single element capable of being used for so many purposes in only one form. For any of this to be remotely possible, vibranium must have been engineered by some of the best chemists and physicists in the world. At an atomic level, vibranium is unique in its ability to form a variety of exotic bonds (metallic, covalent, ionic, etc.) to accommodate its role in so many different forms and functions. This chemical versatility would allow very specific nanoscaled structures to be easily synthesized in conveniently high yields.

The first law of thermodynamics states energy cannot be created or destroyed. With that in mind, vibranium probably can't indefinitely absorb kinetic energy without storing it in another form. This is partly rationalized by Shuri's design of T'Challa's panther habit in *Black Panther*. Whenever T'Challa receives any mechanical stress, it can be absorbed, stored, and released from his panther habit as a purple shock wave. One way to explain this is to consider that the exotic nature of vibranium allows it to form piezoelectric crystal lattices that can be woven into nanite textiles. This unique piezoelectric vibranium weave can transfer kinetic energy into electricity, which can build a charge in a capacitor woven in his habit (see Black Widow's Bite) and be released as a sonic shock wave (see Thor's Lightning or Hulk's Thunder Clap). How would this piezoelectricity work?

We assume vibranium can form a crystal lattice with negatively charged elements that find balance in electric charge within a given unit cell. This unit cell is the repetitive building block of the vibranium, which repeats itself in three dimensions to make up the entire material. In its natural crystalline form, it has a net zero charge across a unit cell, and thus across the entire material. When deformed, however, its crystal lattice squishes to rearrange the charge, causing a potential difference between two sides of the crystal, which can conduct an electrical current.

What about forms of vibranium that are indestructible by mechanical and thermal stressors, like Captain America's shield? If vibranium can form composite crystal lattices with other negatively charged elements, it probably also forms complex nanotubes like carbon. In this example, vibranium can create hexahedral covalent bonds with adjacent molecules, which can be rolled into tubes to create structures that by themselves are more powerful than the arrangement of adjacent vibranium atoms in metallic bonds (see Forging Stormbreaker). This is like taking a thousand popsicle sticks to make a tower that has trusses and a thoughtful infrastructure or taking a thousand popsicle sticks and balancing them vertically one atop the other. Clearly, one structure will be able to handle strain better than the other despite both having the same number of parts made from the same material.

THE SCIENCE OF REAL LIFE

Is there such a thing as real vibranium? For accuracy's sake, there is a smart carbon fiber composite with the name vibranium being used by Hyperloop Transportation Technologies, but it seems to be an attention-seeking PR stunt (Elon Musk strikes again!). That aside, the discovery of piezoelectricity goes as far back as 1880 when Pierre and Jack Curie found the voltage potential generated in quartz. Shortly after, Gabriel Lippmann discovered that the reverse effect could also be found; if an electric current was run through a piezoelectric material, it could induce deformation in the material. Over a century later, piezoelectric materials have been integrated in a variety of commercial goods ranging from children's light-up shoes to ultrasound. Studies carried out in Japan by Dr. Eiichi Fukada in the 1960s have even found that the natural organization of collagen in bone can have piezoelectric effects in response to mechanical pressure. In the 1970s several groups went on to show that bone piezoelectricity was able to affect bone resorption and growth, making it a sensor to strengthen bone. Some metals, despite being piezoelectric, do not retain the ability to form alloys or strong metallic bonds (e.g., lead titanate, $PbTiO_3$, is a white powder). Furthermore, it would be difficult to engineer a crystal that could survive the mechanical duress that T'Challa experiences in his battles as Black Panther.

In the MCU, vibranium is described as an incredible material that is "stronger than steel and a third of its weight." This is impressive—until you realize that carbon-derived graphene is about two hundred times stronger than steel, six times lighter, and can conduct an electric current better than copper. Graphene is an atom thick and is made out of the hexagonal arrangement of carbon grid. However, it would never display the ability to absorb vibration and redistribute that kinetic energy perfectly. This would assume that if a material was deformed, it would be able to return to its original shape with a similar force. This property can be attributed to the elasticity of a material and is seen in polymers such as rubber. These polymers form long chains of hydrocarbons that are very disordered and tangled in their resting state but are cross-linked to one another through covalent bonds. In rubber, when this elastic material is deformed, it forces these long hydrocarbon chains to align in a linear array, which will return to its original conformation once the mechanical strain is removed. However, the inherent molecular structure of an elastic material like rubber is intrinsically antithetical to the strength of a metal. Metals, unlike rubber, rely on forming crystal lattices where a tremendous amount of order organizes positively charged cations of a given metal and distributes their electrons like atomic glue (see Forging Stormbreaker). Our ability to create a variety of new materials with new properties often comes from our ability to mix those materials with other elements and induce nanostructures within them (see Gamora's Sword, Godslayer). Perhaps, though, I may just not be as intelligent and resourceful as the scientists in Wakanda.

GLOSSARY OF MARVEL TERMS

Abilisk
A multi-limbed alien that is attracted to Sovereign's Anulax batteries.

Aero-Rigs
Rocket packs designed by Rocket. The Guardians of the Galaxy use them to fly on different planets.

The Ancient One
The residing Sorcerer Supreme; defender of Earth and mentor to Dr. Strange.

arc reactor
The power source created by Tony Stark to supply his Iron Man suits with energy.

Bifrost Bridge
An Asgardian rainbow bridge that is used to teleport Asgardians across the Nine Realms, under the watchful eye of Heimdall.

Celestials
The oldest and most powerful humanoid beings in the universe.

Chitauri
An alien race recruited by Thanos and Loki to invade Earth. They fought against the Avengers in the Battle of New York.

Ego
A Celestial consisting of a floating brain that can manipulate all forms of matter that created a shell for itself as a living planet.

Extremis
A secret experimental formula invented by Aldrich Killian that can grant its user exothermic abilities, tissue regeneration, and super strength.

Eye of Agamotto
See *Time Stone*. The green Infinity Stone (Time Stone) discovered by the first Sorcerer Supreme, Agamotto.

Flora colossus
A race of treelike beings of which the last known survivor is Groot of the Guardians of the Galaxy.

Godslayer
Gamora's collapsible sword, which is made from extremely dense materials that are very heavy but offset by an energy core hilt that balances the sword evenly.

gravity mine
A weapon used by Star-Lord that can suck in a group of enemies or a specific target with incredible force.

Heart-Shaped Herb
The secret herb kept in the groves of Mount Bashenga that can grant the power of Black Panther to those who ingest it.

HYDRA
A secret authoritarian terrorist group on Earth with aims toward world domination.

Infinity Gauntlet
A glove made from Asgardian uru metal capable of wielding all five Infinity Stones.

Iron Legion
A collection of autopiloted Iron Man suits meant to protect Earth's population from secondary threats.

Iron Man Mark I–L
A series of exosuits developed by Tony Stark with different designs for a variety of functions.

Kree
The blue-skinned alien race that are the sworn enemies of the Skrulls. Notable members are Ronan the Accuser and Korath the Pursuer.

Mind Stone
The yellow Infinity Stone that was wielded in Loki's Scepter to control the minds of others and afterward used to create Ultron and Vision.

Mjolnir and Stormbreaker
Asgardian weapons forged from uru, which can be enchanted to allow their user to travel across space and fly.

Morag
A planet that is home to the temple that stores the Orb of Power and the Power Stone.

nanites
Microscopic nanotechnological machines that can amalgamate to create textiles seen in Black Panther's habit, Iron Man's Mark L, and Iron Spider suits.

Nidavellir
A neutron star forge used by dwarves to create Asgardian weapons.

phasing
A power that Ava Starr (Ghost) has that allows her to travel through physical barriers and across time and space.

Power Stone/Orb of Power
The purple Infinity Stone that was wielded by Ronan the Accuser and the Guardians of the Galaxy before it was lost to Thanos.

Pym Particle
A particle invented by Henry Pym, which can allow its user to shrink in size.

Quantum Realm
A hidden dimension that exists within subatomic space where the rules of physics change.

Quantum Vehicle
A device used by Henry Pym to travel to the Quantum Realm to find his wife, Janet van Dyne.

Reality Stone

The red Infinity Stone that was used to manipulate the fabric of reality.

Redwing

The vertical takeoff and landing surveillance drone that Falcon uses for both battle and reconnaissance.

S.H.I.E.L.D.

Strategic Homeland Intervention, Enforcement and Logistics Division. A secret organization operating above national governments to monitor the activity of metahumans and related technologies on Earth.

Skrulls

A green shapeshifting alien race who are the sworn enemies of the Kree.

Space Stone/Tesseract

The blue Infinity Stone that was used by the Red Skull to fuel HYDRA's weapons. It can be used to make portals such as the one that was used for the invasion of New York City.

Spidey sense

One of Spider-Man's powers, which allows him to have heightened senses in dangerous circumstances.

Super Soldier Serum

A mysterious formula made by Abraham Erskine during the Second World War that was used on Steve Rogers (Captain America) and Johann Schmidt (Red Skull).

Thanos's Snap/The Snap
The moment when Thanos used all five Infinity Stones to wipe out half of the universe's sentient life.

Time Stone/Eye of Agamotto
The green Infinity Stone that was used by Dr. Strange and Thanos to control the flow of time.

Titan
The home planet of Thanos, once a utopia before it suffered a cataclysm due to overpopulation.

Ultron
A sentient artificial intelligence created by Tony Stark and Bruce Banner using the Mind Stone. It turned on the Avengers and believes the extinction of humankind is the only way to have true peace.

uru
An Asgardian ore that requires the power of a dying star to forge legendary weapons.

vibranium
A Wakandan ore derived from a fallen meteor. It is perfectly elastic and capable of deflecting kinetic energy completely.

Vision
A sentient artificial intelligence that fused with J.A.R.V.I.S., a benevolent artificial intelligence that believes in the value of humankind.

Wakanda
A hidden African nation-state of advanced technology based on their use of vibranium. Its sworn protector is the Black Panther.

web-shooters
A special wrist-mounted device that allows Spider-Man to shoot spider silks with which he can swing across the Manhattan skyline.

web wings
An addition to Spider-Man's suit, designed by Tony Stark, that allows him to cover more distance with a pseudo flight suit.

Widow's Bite
An electrical tase that is integrated into all of Black Widow's weapons and her bodysuit.

Winter Soldier
A Super Soldier assassin, Bucky Barnes, manipulated by HYDRA, who erased his memory and put him into cryostasis after each mission.

GLOSSARY OF SCIENCE TERMS

Bessel beams
These are acoustic, electromagnetic or gravitational waves defined by a Bessel function that have tractor beam like properties.

biomimetics
Technology that emulates a biological morphology or function.

biosequestration
The process in which a living organism can collect or concentrate a specific element(s) within its tissues.

black hole
A singularity of an incredibly dense point in space that is often the result of a dying star or primordial black holes that occurred during the Big Bang.

capacitor
Two plates that can collect a potential difference and store a large electric charge.

carbon nanotubes
A nanostructure composed of the tubelike arrangement of a hexagonal sheet of carbon atoms. They can imbue a material with great tensile strength.

cerebral cortex
The outer layers of corrugated tissue on the surface of the brain that have a large surface area; where several higher-order thinking processes occur in vertebrate mammals.

ciliary muscles
These muscles control the positioning of the lens in a human eye.

closed timelike curve (CTC)
A time loop where an object's world line loops back in time only to reappear back at the same point in time.

CRISPR
A technology that allows for the surgical manipulation of specific sequences of DNA.

cryobiology
The study of how life survives in the cold.

cryostasis
The phenomenon of how life can remain preserved at low temperatures.

cybernetics
The study of machine–life interfaces.

De Broglie wavelength
A function used to describe the wave properties of matter.

DNA methylation
A covalent modification that can be made to different parts of a gene that can control the function and/or transcription of the gene.

double-stranded DNA breaks
Incidents where the phosphate backbone of both strands that make up the DNA double helix break.

Einstein-Rosen bridge
A theory that suggests the existence of wormholes that can allow matter to travel between two points with aberrant time-space continuums.

electromyography (EMG)
The recording of electrical activity from a muscle.

epidemiology
The study of how diseases are caused, distributed, and controlled within populations.

epigenetics
The field of study that examines how environmental variables can shape genomic responses through various plastic molecular mechanisms.

erythropoietin
An important regulator of hemoglobin production in red platelet cells.

fractal cosmology
The cosmological theory that the universe is a nested cluster of galaxies at every scale of existence.

gamma radiation
A high-energy form of electromagnetic radiation capable of stripping electrons from other molecules and atoms, creating oxidative radicals.

gene(s)/genome
The sequence of coded chemical information that can be used as a blueprint to make proteins that carry out specific biological functions.

gene editing
The process of taking a sequence of the genetic code and modifying parts of it artificially.

gravitational waves
The gravitational force that travels through space-time when objects with incredibly large mass move.

holography
The use of a scattered wave and reference wave interacting with each other to create a three-dimensional interference pattern that can be used to recreate physical properties of the object that scattered the original wave.

lightning leader
An electrostatic discharge released from a lightning cloud.

melanocytes
Epithelial cells that are responsible for depositing pigment into the skin.

metallic bonds
A unique bond that can exist within metallic elements that organize themselves into a crystal lattice while using disassociated outer valence electrons to form a glue that holds together the crystal lattice.

metallurgy
The study of metals, their production, and purification.

myostatin
A protein responsible for repressing typical muscle growth. Inhibition of myostatin can cause significant muscle growth.

nanocarriers
A nanomaterial that is being used to carry another substance such as a drug.

nanotechnology
The study of objects and technologies that are smaller than 100 nanometers.

neuroepigenetics
The study of epigenetic mechanisms in the brain.

nuclear fission
A process that involves the splitting of an atom's nucleus; leads to the release of large amounts of energy.

occipital cortex
The cortical layers that are responsible for processing visual information that is sent through the optical tracts from the eyes.

optical trapping
The use of two or more sources of electromagnetic radiation to exert a trapping force on a small object; also called laser tweezing.

optogenetics
The transgenic use of light-sensitive proteins from algae that can be triggered to cause a neuron(s) to fire.

oxidative radicals
Charged atoms or molecules capable of wreaking molecular havoc on biological molecules such as nucleic acids, proteins, or lipids.

parthenogenesis
Reproduction from female sex cells without fertilization from a male sex cell.

peripheral nerve interface (PNI)
A computational interface that can detect electrical impulses transmitted from nervous tissue.

photonics
The study of photons (light) and their properties.

prefrontal cortex
The frontal area of the brain that is responsible for executive function, emotional processing, and various higher-order thinking processes.

protein
The functionalized coded product of a gene that can carry out various biological processes.

quantum entanglement
The process in which two quantum particles can be linked.

quantum mechanics
The mathematical description of motion of subatomic particles.

quantum superposition
The idea in which a quantum particle can exist in multiple states before its probability wave function collapses to a single observable outcome.

quantum tunneling
The ability of a quantum particle to pass through a physical barrier.

quiet eye
The phenomenon of the observed extra time a skilled athlete takes before carrying out a motor task.

sonic shock wave
The process in which a sound wave travels faster than the speed of sound, causing air to heat up and create a sonic boom.

spaghettification
The phenomenon where matter gets acted upon by such strong gravitational force that it succumbs to being stretched into a thread of single atoms, like a piece of spaghetti.

synapse
The cleft that exists between neurons that allows chemical information to transmit between neurons.

systems neuroscience
The study of neurons and their circuits as whole systems that regulate the nervous system.

telomerase
An enzyme responsible for restoring the nucleotide ends of chromosomes during aging.

time dilation
The process in which gravity can curve the time-space continuum to increase the relative length of time as objects get closer to the speed of light.

transcranial magnetic stimulation (TMS)
A noninvasive means to locally control different parts of the cerebral cortex using powerful electromagnets.

transcription
The process in which a gene is coded into a transient messenger RNA molecule before it is translated into a protein.

vegetative propagation
The process in which the stem-like state of a plant tissue can give rise to necessary tissues to create a new clonal plant.

vitrification
A form of cryopreservation in which a sample goes through an immediate solid state, freezing without the formation of ice crystals.

world line
The four-dimensional path followed by a given object in time-space. It can be used to illustrate the paths of particles through closed timelike curves (CTCs).

wormhole
A theoretical phenomenon that allows for a portal to facilitate travel between two points at faster-than-light travel.

US/METRIC CONVERSION CHART

LENGTH CONVERSIONS

US LENGTH MEASURE	METRIC EQUIVALENT
¼ inch	0.6 centimeters
½ inch	1.2 centimeters
¾ inch	1.9 centimeters
1 inch	2.5 centimeters
1½ inches	3.8 centimeters
1 foot	0.3 meters
1 yard	0.9 meters

WEIGHT CONVERSIONS

US WEIGHT MEASURE	METRIC EQUIVALENT
½ ounce	15 grams
1 ounce	30 grams
2 ounces	60 grams
3 ounces	85 grams
¼ pound (4 ounces)	115 grams
½ pound (8 ounces)	225 grams
¾ pound (12 ounces)	340 grams
1 pound (16 ounces)	454 grams

INDEX

myosin protein, 157–59
spider silk proteins, 103–6
structural proteins, 86, 104
synthetic proteins, 69–70
Psychology, 12–16, 48–52
Pym, Henry, 35–36, 183–86, 214
Pym Particles, 35–36, 183–85, 214

Quake, 82
Quantum entanglement, 185, 214, 225
Quantum mechanics, 176–87, 195, 225
Quantum Realm, 183–87, 214, 215
Quantum superposition, 185–86, 225
Quantum tunneling, 179–82, 225
Quantum Vehicle, 184–85, 214
Quiet eye training, 15–16, 225

Radio waves, 107–11, 200
Radioactivity, 157, 186, 201
Reality, manipulating, 150–54
Reality Stone, 151, 154, 215
Red Skull, 74, 174, 215, 216
Redwing, 57, 107–11, 215
Relativity theory, 136, 174–78, 186. *See also* General relativity
Rhodes, Jim, 94–95, 138
Robotics, 98–102, 206
Robots, 13, 98–102, 138
Rocket propulsion, 95, 111–15
Rocket Raccoon
Aero-Rig, 111–15, 211
animal intelligence, 42–46
nuclear fission, 168
Rogers, Steve, 20–23, 68–69, 74–79, 207, 216
Ronan the Accuser, 134, 168–70, 213, 214

Samuels, David, 82
Savin, Eric, 82
Scarlet Witch, 24–26
Scenarios, stupendous, 155–72
Schrödinger's cat, 186
Schrödinger's equation, 181
Science terms, 219–26
Scorch, 82
Selvig, Erik, 24–25
Senescence, 88–90
Sensory neuroscience, 48–52, 57–62
Shapeshifting, 202–6
S.H.I.E.L.D., 12–14, 74, 82–86, 109, 175–79, 202–6, 215
Shock waves, 119, 123–24, 208, 225
Skrulls, 202–5, 213, 215
Snap of Thanos, 160–64, 216
Solar radiation, 40, 152, 201
Sonic shock wave, 119, 208, 225
Sound waves, 50, 109, 118–24, 225
Space Stone, 161, 174–76, 215
Spaghettification, 167, 177, 225
Spectroscopy, 170–72
Spider cognition, 57–60
Spider grip, 146–50
Spider morphology, 146–50
Spider webs, 59–60, 102–6
Spider-Man
artificial intelligence, 138–42
exosuits, 94–96
gripping abilities, 146–50
nanites, 98–102
spider webs, 59–60, 102–6
Spidey sense, 57–60, 215
wall crawling abilities, 146–50
web wings, 187–91, 217
web-shooters, 102–6, 217
Spider-Man: Homecoming, 57–60, 94–96, 102–6, 138–42, 146–50, 187–91

ABOUT THE AUTHOR

Sebastian Alvarado worked in the biotech industry for four years before pursuing an academic career. He earned a PhD at McGill University in Canada, where he studied the molecular mechanisms that lend plasticity to biological systems. He received an A.P. Giannini Foundation postdoctoral fellowship to continue his research at Stanford University. Currently he is an assistant professor at Queens College of the City University of New York. Sebastian's interest in science began with his wish to understand the genetic basis of the X-Men's superpowers. After realizing the impossibility of attaining those powers himself, he did the next best thing: became a scientist. He is cofounder of a science consulting and communication firm, Thwacke!, aimed at improving the portrayal of science in the entertainment sector.

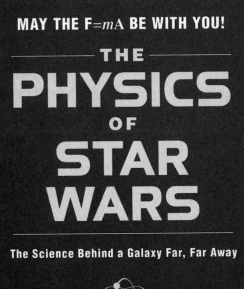